Dedication

This book is dedicated to The College of New Jersey, a shining light in the education of New Jersey's citizens for almost 150 years. That light is still shining because of its dedicated faculty and caring administrators, who include Dr. R. Barbara Gitenstein, President; Dr. Suzanne H. Pasch, Dean of the School of Education; and Dr. Sharon Sherman, Chairperson of the Department of Elementary and Early Childhood Education.

J. F. Silver

ACKNOWLEDGMENTS

Several individuals and publishing houses have been of immeasurable help in the creation of *Geography Curriculum Activities Kit*. Lorna Jean Silver, who received her B.A. in English from Susquehanna University, PA, and her M.A. in English from the Bread Loaf School of English, Middlebury College, VT, edited the entire manuscript and answer key. Her ability to find and address matters needing correction was remarkable.

Mary Salerno, a graduate of The College of New Jersey's School of Business, who is now a free-lance computer and desktop publishing professional, made ready the entire book, including most of the charts, graphs, tables, diagrams, and maps. In the early stages of the production, she was assisted by Alex Pan of The College of New Jersey faculty and his wife, Beatrice.

Special thanks are given to Silver Burdett Ginn Company for permission to use instructional materials from books published by them and authored by James F. Silver—including skillbooks for *The Changing New World, Learning About Latin America, Our Big World, Old World Lands*, and *American Continents*.

GEOGRAPHY CURRICULUM ACTIVITIES KIT

Ready-to-Use Lessons and Skillsheets for Grades 5-12

JAMES F. SILVER

THE CENTER FOR APPLIED RESEARCH IN EDUCATION
Paramus, New Jersey 07652

Library of Congress Cataloging-in-Publication Data

Silver, James F.

The CIP Data is available from the Library of Congress
under LCCN 00-034586.

Printed in the United States of America

10 9 8 7 6 5 4 3 2 1   10 9 8 7 6 5 4 3 2

ISBN 0-13-016119-5 (spiral)   ISBN 0-13-042591-5 (paper)

**THE CENTER FOR APPLIED RESEARCH
IN EDUCATION**
Paramus, New Jersey 07652

http://www.phdirect.com

ABOUT THE AUTHOR

James F. Silver received his B.A. in Social Studies from Montclair University, Montclair, NJ; his M.A. in History from Boston University; his M.A. in Educational Administration from Montclair University; and his Ed.D. in Curriculum and Instruction from Pacific Western University in Los Angeles. Professor Silver also studied the psychology of reading at Temple University in Philadelphia, which led to his New Jersey state certification as a reading specialist.

Dr. Silver's experience includes nine years as an elementary school teacher and principal in Morris County, New Jersey, and more than 35 years in the School of Education at The College of New Jersey, where he is now Professor Emeritus in Residence.

During his educational career, Professor Silver has written numerous teacher's manuals, geography and history skill development books, and achievement tests for Silver Burdett Company. He has also written the two-volume *United States Yesterday and Today* for Ginn and Company. He is author of *Geography Skills Activity Kit*, *World Geography Activities*, *American History Activities*, *The American Continents*, and *Real Life Reading Activities*—all published by The Center for Applied Research in Education, a part of the Prentice-Hall publishing complex. Prior to authoring *American History Activities*, he authored *Environmental Awareness*, published by Kendall/Hunt Company. The last five books cited above were written on the professional level for in-service teachers. His latest book is *New Jersey—Its Past and Present in Geography and History*.

About This Resource

Geography Curriculum Activities Kit is a book designed to help students gain mastery of the skills necessary to properly utilize maps, charts, diagrams, graphs, tables, pictures, and narratives. Thus, for example, when students are faced with a map in a textbook, they will be better able to read the map as it relates to the context of a page or chapter. An example: Students are studying the American Revolution. On a particular page a map is shown. The text may not even refer to the map, or, if it does refer to the map, it makes only a brief reference. A treasure of information and implications is lost if the teacher doesn't direct the students' attention to the map or if the students don't feel comfortable reading the map.

Assume that the map shows the American colonies along the Atlantic coast, the Atlantic Ocean, and Great Britain. By the teacher asking students to use the map's scale of miles they can determine the approximate distance across the Atlantic Ocean. Then, the teacher can ask an inference question such as, "Why was it difficult for the British to fight a war some 3500 miles away from their country?" or "If a sailing warship had favorable winds and could sail 200 miles in 24 hours, how many days would it take to transport soldiers and supplies to America?" or "How would the time and effort affect the British war effort?" (Geography and wars are inseparable—no war was ever fought that wasn't affected by geographical conditions.)

Every student activity in the book offers numerous opportunities for skill development, not only with maps, but also with several other visuals such as graphs and pictures. Students have numerous opportunities to read pictures, for example, for main ideas, details, and inferences. To say, "A picture is worth a thousand words," is very true—provided that students have been trained to read pictures. One of the pictures in the book shows a desert scene that requires students to answer 16 detail questions, 9 inference questions, and one question concerned with determining the main idea of the picture. It's conceivable that it would take an entire page to tell what the picture tells. And, in some ways the picture would still be more effective.

Every student activity in *Geography Curriculum Activities Kit* develops at least one major geography/social studies skill. The transfer and application of the developed skills to contextual situations is a short jump for the simple reason that the skills are developed in contextual situations. Additionally, students who have read, studied, and reacted to the skill-development opportunities are almost certain to improve their performance on standardized achievement tests.

Instructor's Pages

Instructor's pages are an important part of *Geography Curriculum Activities Kit*. They are designed to supply background information relative to the subject matter and skills students will be studying. For example, in Section 2 of the book there is an instructional page titled "Place-Name Geography." The page explains why place-name geography is important and then suggests how it might be taught throughout the semester. Another example, Section 5 contains an instructor's page titled "Latitude As a Measuring Device." This page explains how to use latitude as a tool for measuring distance; it also contains a diagram and a table of distances that tells how many miles a degree of latitude is equivalent to on various lines. The accompanying student pages provide numerous opportunities to determine latitudinal locations and distance measurement.

Many of the text's 66 instructor's pages can be used as information background for students. The pages are written in a style and on a reading level that most students can comprehend. "Altitude—Its Measurement and Significance," is such a page. Its content is enhanced by an accompanying student activity page that contains a special diagram and requires 16 responses.

Skill Development

The text's student activity pages attend to developing skills in reading and making graphics. Few people forget the basic addition, subtraction, multiplication, and division skills and facts. Why? The answer is that, through introduction, constant repetition, and extension, arithmetic facts and skills become part of a person's store of information and procedures. Likewise, the more one utilizes skills in map reading, table preparation and interpretation, graph making and reading, etc., the more lasting the memory and the greater the expertise.

Every student page contains at least one major skill objective as well as several components of that skill. For example, the major topic on page 73 is told in the title of the page, "Completing and Understanding a Color Elevation Map." The major skill is reading altitude on a contour map. The sub-skills are using the key of the map to interpret altitude, observing proximity of contour lines to determine steepness, using altitude to determine sources and flows of rivers, and using color to determine altitude. Another activity, page 75, "Climbing Bell Mountain," uses these same skills in a different and more sophisticated setting. Additional elements are added—contour interval, determining the exact altitude of objects, determining distance on a contour map, and completing a table listing of elevations. Then, to take the gradual sophistication of reading altitude further, students complete page 76, "Reading a United States Topographical Map." In this activity students are asked to "read" more complex contour lines and to interpret more abstract symbols.

The reading, interpreting, and making of graphs is another example of progressive growth in skill development in *Geography Curriculum Activities Kit*. Picture graphs, bar graphs, circle graphs, square graphs, and line graphs are all given considerable attention. Five instructor's pages and ten student activity pages are included in Section 5, "Graphs." The activities employ topics as divergent as cotton production, the highest elevations on continents, frequency of tornadoes, and population growth to capture student interest.

The completion of all activities will also increase students' reading comprehension and study skills. Frequent questions are asked that require finding specific details in the text. In many cases students are asked to reread and circle phrases and sentences that convey particular understandings. For example, in the page titled "What Happened to Riverside?" students are asked to "Circle the two sentences in the story that tell why the run-off from the dump killed fish in the river."

Following is a list of the number of graphics utilized by this text in various categories: Maps, 47; Graphs, 25; Tables, 14; Diagrams, 60; Pictures, 56.

Table of Contents

ABOUT THIS RESOURCE vi

Section 1
Basic Map Reading Skills • 1

1-1   Creating a Map from Symbols Obtained from a Key (Instructor) 2

1-2   Drawing a Map 3

1-3   Developing Some Important Geographic Terms (Instructor) 4

1-4   A Map with Natural and Human-made Features 5

1-5   Reading and Interpreting Cross-Section and Aerial Diagrams (Instructor) 6

1-6   Identifying Land, Water, and Sky Features 7

1-7   Recognizing Boundaries and Routes of Travel (Instructor) 8

1-8   Map Symbols for Things That Cannot Be Seen 9

1-9   Help In Teaching Directions (Instructor) 10

1-10   Using Directions to Locate Cities 11

1-11   Completing a Direction Maze 12

1-12   Finding Your Way Around a Town 13

1-13   Determining Distances with a Scale of Miles (Instructor) 14

1-14   Using a Scale of Miles to Find Distances on Maps 15

1-15   Reading Road Maps (Instructor) 16

1-16   Finding Your Way on a Road Map 17

1-17   Direction and Distance Around Wildwood Lake 18

1-18   Using an Index to Locate Places 19

Section 2
Developing a Sense of Place • 21

2-1   Place Name Geography (Instructor) 22

2-2   United States Map 23

2-3   Crossword Puzzle of the United States I 24

2-4   Crossword Puzzle of the United States II 25

2-5   All About Rivers (Instructor) 26

2-6   The Mississippi Valley 27

2-7   Wordsearch on East-of-the-Mississippi River Capitals 28

2-8   The Eastern States and Their Capitals 29

2-9   Wordsearch on West-of-the-Mississippi River Capitals 30

2-10   The Western States and Their Capitals 31

2-11   The World's Major Land and Water Divisions (Instructor) 32
2-12   The World's Continents and Oceans 33

Section 3
World Water Passages • 35

3-1   The Panama Canal: Some Basic Information (Instructor) 36
3-2   Sail Through the Panama Canal 37
3-3   The Suez Canal: Some Basic Information (Instructor) 38
3-4   Sail Through the Suez Canal 39
3-5   The Dardanelles, Sea of Marmara, Bosporus: Some Basic Information (Instructor) 40
3-6   Sail From the Mediterranean Sea to the Black Sea 41
3-7   The Strait of Gibraltar: Some Basic Information (Instructor) 42
3-8   Sail Through the Strait of Gibraltar 43
3-9   The Strait of Magellan: Some Basic Information (Instructor) 44
3-10   Sail Through the Strait of Magellan 45
3-11   Sail Through the Bering Strait 46

Section 4
Maps That Show a Pattern • 47

4-1   Pattern Maps (Instructor) 48
4-2   Traveling by Road and Railroad 49
4-3   Using Two Maps to Find Answers to Questions 50
4-4   A Pattern Map of the World's Deserts 51
4-5   An Altitude Pattern Map of 19 Western States 52
4-6   A Pattern Map of Foreign Visitors to the United States 53
4-7   The Leading Barley-Producing States 54
4-8   Coastlines of the United States 55

Section 5
The Earth's Grid • 57

5-1   Latitude As a Measurement Device (Instructor) 58
5-2   Constructing a Latitude Diagram 59
5-3   Finding Distances Between Places on Lines of Latitude 60
5-4   Determining Latitude and Distance on a World Map 61
5-5   Special Lines of Latitude That Define Earth's Zones (Instructor) 62
5-6   Understanding the World's Climate Zones 63
5-7   Longitude as a Measurement Device (Instructor) 64
5-8   Locating Places and Measuring Distance on Lines of Longitude 65

5-9 Using Latitude and Longitude to Find Locations and Distances 66

5-10 Longitude Helps Us to Tell Time 67

Section 6
Understanding Altitude • 69

6-1 Altitude: Its Measurement and Significance (Instructor) 70

6-2 Understanding Altitude 71

6-3 The Use of Color To Show Elevation (Instructor) 72

6-4 Completing and Understanding a Color-Elevation Map 73

6-5 Understanding Contour Maps (Instructor) 74

6-6 Climbing Bell Mountain 75

6-7 Reading a United States Topographical Map 76

Section 7
Global Maps and Polar Maps • 77

7-1 Map Projections: Some Pro and Con (Instructor) 78

7-2 Globe Maps and Lines of Latitude 79

7-3 Working With Longitude I 80

7-4 Working With Longitude II 81

7-5 Latitude and Longitude on a Globe Map 82

7-6 Polar Maps (Instructor) 83

7-7 Working with Polar Maps I 84

7-8 Working with Polar Maps II 85

Section 8
Graphs • 87

8-1 Reading Picture Graphs (Instructor) 88

8-2 A Picture Graph Shows a Farmer's Apple Harvest 89

8-3 A Picture Graph of a Family's Jam-Making Activities 90

8-4 A Picture Graph Showing Cotton Production in the United States 91

8-5 Reading Bar Graphs (Instructor) 92

8-6 The United States' Greatest Oil-Producing States 93

8-7 Comparing the Highest Points of Continents 94

8-8 Making and Reading Circle Graphs (Instructor) 95

8-9 Comparing Quantities and Sizes On Circle Graphs 96

8-10 Making and Reading Square Graphs and Single-Bar Graphs (Instructor) 97

8-11 A Square Graph Compares South America's Countries 98

8-12 Two Single-Bar Graphs: The Great Lakes and Fishing 99
8-13 Making and Reading Line Graphs (Instructor) 100
8-14 Understanding Tornadoes 101
8-15 Completing a Line Graph of Population Growth 102

Section 9
Soil Erosion and Prevention of Erosion • 103

9-1   All About Soil (Instructor) 104
9-2   What Happened to Riverside? I 105
9-3   What Happened to Riverside? II 106
9-4   Soil Erosion: Construction, Roadsides, and Strip-Mining I 107
9-5   Soil Erosion: Construction, Roadsides, and Strip-Mining II 108
9-6   Causes of Soil Erosion: Overgrazing, Bare Fields I 109
9-7   Causes of Soil Erosion: Overgrazing, Bare Fields II 110
9-8   The Importance of Sedimentation (Instructor) 111
9-9   Splash Erosion: Activity (Instructor) 112
9-10 Ways of Protecting Soil I 113
9-11 Ways of Protecting Soil II 114
9-12 Lesson Planning: Developing Content, Skills, and Values (Instructor) 115
9-13 Lesson Planning: Causes and Remedies for Soil Erosion (Instructor) 116
9-14 Lesson Planning: Instructions To Students and Outcomes (Instructor) 117
9-15 Lesson Planning: Example Farm Design (Instructor) 118

Section 10
Water: Its Sources and Its Pollution • 119

10-1   All About Water (Instructor) 120
10-2   "Water" Is an Important Word (Instructor) 121
10-3   Water: Too Precious to Waste (Instructor) 122
10-4   Understanding Watersheds 123
10-5   More About Watersheds I (Instructor) 124
10-6   More About Watersheds II 125
10-7   The Water Cycle: An Explanation (Instructor) 126
10-8   Understanding the Water Cycle 127
10-9   Water from Underground Sources I (Instructor) 128
10-10 Water from Underground Sources II 129
10-11 Sources of Fresh Water: Surface I 130
10-12 Sources of Fresh Water: Surface II 131
10-13 Water Pollution: Spilled Oil I (Instructor) 132
10-14 Water Pollution: Spilled Oil II 133

10-15 Oil Spill in the Shetlands: An Environmental Disaster I 134

10-16 Oil Spill in the Shetlands: An Environmental Disaster II 135

10-17 Thermal Pollution: What Is It? How Can It Be Controlled? 136

Section 11
Air: What Is It; How Is It Polluted? • 137

11-1 All About Air I (Instructor) 138

11-2 All About Air II 139

11-3 Demonstration: Proving That Air Has Weight (Instructor) 140

11-4 Understanding the Atmosphere 141

11-5 What Is Air Pollution? I (Instructor) 142

11-6 What Is Air Pollution? II 143

11-7 The Effect of Acid Rain on Green Plants (Instructor) 144

11-8 Acid Lakes: What Can Be Done About Them? 145

11-9 Air Pollution: Industry and Power Plants I 146

11-10 Air Pollution: Industry and Power Plants II 147

11-11 Polluted Air Is Expensive I 148

11-12 Polluted Air Is Expensive II 149

11-13 A Polluted Air Disaster 150

Section 12
Wildlife in Geography and the Environment • 151

12-1 Wildlife in the News: Eagles I (Instructor) 152

12-2 Wildlife in the News: Eagles II 153

12-3 Trees: What We Get From Them I 154

12-4 Trees: What We Get From Them II 155

12-5 Redwood Trees Make a Comeback I (Instructor) 156

12-6 Redwood Trees Make a Comeback II 157

12-7 "Space" and How It Affects Animal Populations I (Instructor) 158

12-8 "Space" and How It Affects Animal Populations II 159

12-9 All About Whales I (Instructor) 160

12-10 All About Whales II 161

12-11 Wildlife Word Search 162

Section 13
Earth's Natural Wonders • 163

13-1 All About Earthquakes I (Instructor) 164

13-2 All About Earthquakes II 165

13-3   Volcanoes: Earth's Chimneys (Instructor) 166
13-4   The Making of a Volcano 167
13-5   All About Glaciers I (Instructor) 168
13-6   All About Glaciers II 169
13-7   Understanding Icebergs and Ice Sheets 170
13-8   Understanding Hurricanes 171
13-9   Geysers: Natural Water Fountains 172
13-10 Caves: How They Are Made and How They Were Useful in Early Times 173

Section 14
Learning from Pictures • 175

14-1   Developing Picture-Reading Skills (Instructor) 176
14-2   Working Together in Switzerland 177
14-3   In the Netherlands Dikes Hold Back the Ocean 178
14-4   A Mexican Village 179
14-5   Animal Friends and Enemies in Australia 180
14-6   Life in a Desert 181
14-7   Fishing in the United States 182
14-8   Life on the Plains in Early Times 183
14-9   Nature and Humans Change the Face of Earth 184

Section 15
Earth, Sun, and Moon Relationships • 185

15-1   Day and Night: What Causes Them? 186
15-2   The Moon: Some Factual Information (Instructor) 187
15-3   Remembering Moon Facts 188
15-4   The Sun and It's Relationship to Earth (Instructor) 189
15-5   Seasons North and South of the Equator 190
15-6   All About Compasses and Magnetic North (Instructor) 191
15-7   Understanding the Compass: Direction and Time 192

Section 16
Activities • 193

16-1   Geography Jigsaw Puzzle (Instructor) 194
16-2   Crossword Countries of the World 195
16-3   Geography Dioramas (Instructor) 196
16-4   Research: Questions and Answers I (Instructor) 197
16-5   Research: Questions and Answers II (Instructor) 198
16-6   Geography Bingo I (Instructor) 199

16-7   Geography Bingo II (Instructor) 200

16-8   Geography Bingo III (Instructor) 201

16-9   Geography Bingo IV (Instructor) 202

16-10 Latitude and Longitude Treasure Hunt 203

16-11 "TV Filmstrip" in a Box (Instructor) 204

16-12 Unscramble the Names of Countries and Capitals 205

16-13 Geography Wordsearch 206

16-14 Environmental Riddles 207

16-15 Environmental Poetry 208

Section 17
Outline Maps • 209

17–1   North America (labeled) 210

17–2   North America (unlabeled) 211

17–3   USA (labeled) 212

17–4   USA (unlabeled) 213

17–5   Middle America (labeled) 214

17–6   Middle America (unlabeled) 215

17–7   South America (labeled) 216

17–8   South America (unlabeled) 217

17–9   South Asia (labeled) 218

17–10 South Asia (unlabeled) 219

17–11 Nations of the Former USSR (labeled) 220

17–12 Nations of the Former USSR (unlabeled) 221

17–13 Australia and New Zealand (labeled) 222

17–14 Australia and New Zealand (unlabeled) 223

17–15 Africa (labeled) 224

17–16 Africa (unlabeled) 225

17–17 Europe (labeled) 226

17–18 Europe (unlabeled) 227

17–19 North Pole (labeled) 228

17–20 North Pole (unlabeled) 229

17–21 South Pole (labeled) 230

17–22 South Pole (unlabeled) 231

Section 18
Answer Key • 233

SECTION 1

BASIC MAP-READING SKILLS

1-1   Creating a Map from Symbols Obtained from a Key (Instructor) 2

1-2   Drawing a Map 3

1-3   Developing Some Important Geographic Terms (Instructor).........................4

1-4   A Map with Natural and Human-Made Features 5

1-5   Reading and Interpreting Cross-Section and Aerial Diagrams (Instructor) 6

1-6   Identifying Land, Water, and Sky Features 7

1-7   Recognizing Boundaries and Routes of Travel (Instructor)........................8

1-8   Map Symbols for Things That Cannot Be Seen 9

1-9   Help In Teaching Directions (Instructor) 10

1-10 Using Directions to Locate Cities 11

1-11 Completing a Direction Maze 12

1-12 Finding Your Way Around a Town 13

1-13 Determining Distances with a Scale of Miles (Instructor) 14

1-14 Using a Scale of Miles to Find Distances on Maps 15

1-15 Reading Road Maps (Instructor) 16

1-16 Finding Your Way on a Road Map 17

1-17 Direction and Distance Around Wildwood Lake 18

1-18 Using an Index to Locate Places 19

CREATING A MAP FROM SYMBOLS OBTAINED FROM A KEY

Background

Young learners have had considerable experience attaching meaning to symbols. Most primary-grade children understand that word symbols represent real things: *tree, house, dog*, etc. They may realize that the flag in the front of the classroom stands for the United States. When they are riding in an automobile and see a sign with the silhouette of a deer, they probably realize that deer are near and that the driver should be on the lookout for them. And, they know enough to stay on a path that shows a bicycle and, perhaps, an arrow indicating direction. More than likely they probably have been told that when a stop light shows red they should stop, and if it shows green they may go.

Maps, of course, are loaded with symbols. Because a map is a representation of a part of the earth's surface, symbols of a great variety are employed. This means that if one wants to read a map one must recognize the symbols used. Most often the symbols on beginning maps are shown and identified in the *key* of a map. However, after learners have had sufficient experience, the maps they use may show and indicate the meaning of only special or unusual symbols.

Beginners' maps utilize pictorial symbols, but the time comes when symbols of a more abstract nature are used. Following are some of the elements of maps that are shown by semi-abstract and abstract symbols.

- coastlines	- lakes	- roads
- peninsulas	- mountains	- bicycle
- islands	- bays, gulfs	paths
- rivers,	- railroad	- boundaries
streams	lines	- bridges
- airports	- parks	- communities

An examination of the listed symbols shows both natural and human-made features. For example, rivers are natural, but dams are not. Often, natural and unnatural features are shown in combination, as in the case of a river that is crossed via a tunnel or a bridge.

Another important aspect of map reading is that when we look at a map we are looking *down* as though we are in an airplane above the earth. If we take pictures of what we see, we have what is called an *aerial* view. Then, through the use of symbols, we can make a map to represent what the photograph shows pictorially. For example, we could use a line for a road or river and a dot for a city.

Student Involvement

1. The activity on the facing page requires students to complete a map of EREHWON (NOWHERE spelled in reverse). After students have completed their maps, it would be encouraging and motivating to display their maps on a bulletin board.

2. There are some implied understandings of which students may not be aware. For example, an apple orchard is a planned agricultural endeavor. Therefore, students should draw the trees in rows. However, forests are unplanned: nature does not set trees in rows, so there is no particular order.

3. The activity may be extended and become more meaningful if students are directed to fill the spaces between segments E, J, L, K, and M with country-type roads. For example, parallel lines could be drawn in the spaces and filled with a suitable symbol (•⋮•⋮•). The new symbol could be added to the key of the map and labeled *Country road*.

Name: _____ Date: _____

DRAWING A MAP

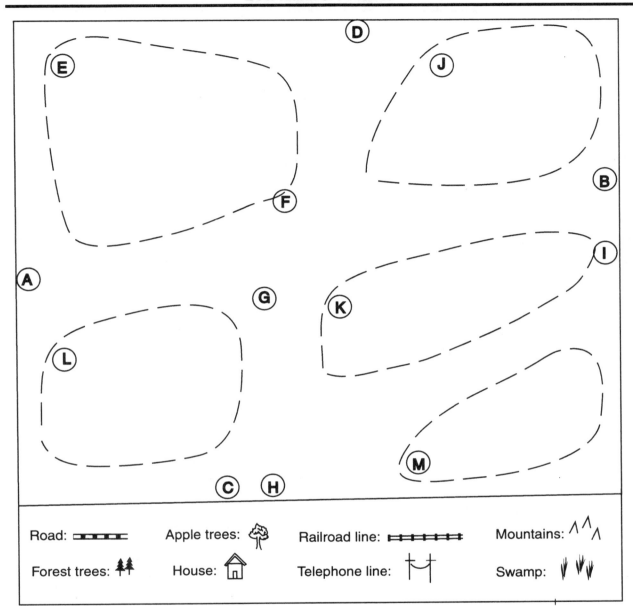

To Do:

Use the symbols shown in the key of the map to complete the map of EREHWON. Use a ruler where straight lines are needed.

1. Draw a straight road from A to B.

2. Draw another straight road from C to D.

3. Find the part of the map shown by letter E, and fill it with rows of trees to show an apple orchard.

4. Draw a house at letter F and another house at letter G.

5. Draw a telephone line along the upper side of the road from A to B.

6. Draw a straight railroad line from H to I.

7. Draw mountains in section J.

8. Draw a swamp in section K.

9. Draw forests in sections L and M.

10. *Lightly* color the apple orchard red, the forests green, and the swamp blue.

DEVELOPING SOME IMPORTANT GEOGRAPHIC TERMS

Background

The activity on the facing page will reinforce what students may know, or bring their attention to vocabulary they do not know or of which they are unsure. And, as in the previous activity, significant emphasis is placed on the recognition of geography symbols.

Geography vocabulary is important; many place-names throughout the world include geographical terms. It may be of advantage to point out on a map uses of the terms used in the activity: Bering Strait, English Channel, Baja Peninsula of Lower California, Bay of Biscay, Gulf of Mexico, Cape Horn, Cape Cod.

Point out the significance of some of the terms. For example, the Bering Strait, which was once an isthmus, enabled Asiatics to cross into North America. The English Channel offered the island of Great Britain protection from invaders. Gulfs and bays provide safe harbors for ships. Cape Hatteras, jutting into the Atlantic Ocean, was the site of a lighthouse that protected ships at sea.

Suggestions for Teaching

1. The maps and definitions that follow may be made into a transparency and prove useful in the instructor's explanations.

2. After students have completed the activity, it would be useful for them to make up a map showing all the terms explained on this page. In addition, they should try to include the symbols shown in the key to the map.

Students also should give names to the elements they draw. Suggest that they use their imaginations in the drawings and in the names they apply to features. For example, it should be easy to see why *Big Foot Peninsula* is so called and how *Cape Toe*'s name was derived.

To stimulate students' imaginations further, make some transparencies of the following: *Fish Hook Island*, shaped like a hook; *Circle Bay* named so because it is almost a closed circle; and *Crescent Mountain Range* a range shaped like a crescent; and, finally, *Hour Glass Channel*, named so because it has the shape of an hour glass.

Peninsula: Land that has three sides and juts into a body of water such as an ocean, sea, or lake.

Bay or Gulf: A part of an ocean, sea, or lake that extends inland. Generally, bays are smaller than gulfs.

Straits/channels: Narrow water passages between two pieces of land.

Isthmus: A narrow strip of land that connects two larger pieces of land.

Cape: A piece of land, generally pointed but often rounded, that juts into a body of water. Capes resemble peninsulas, but they are often smaller.

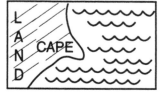

A MAP WITH NATURAL AND HUMAN-MADE FEATURES

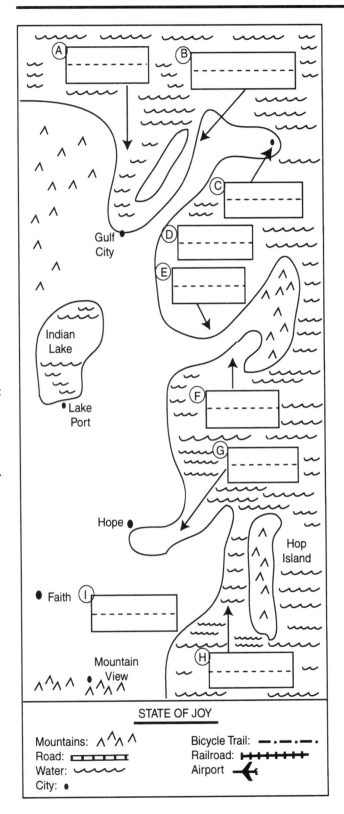

Completing the Map

The map is a make-believe map of the State of Joy. Your task will be to complete the map according to the directions that follow.

The verb *label*, as used below, means to print names of places in the lettered boxes. When you label the places, try to print carefully and small.

To Do:

1. Label the following on the map.

   - Ⓐ Gulf of Tango
   - Ⓑ Big Foot Peninsula
   - Ⓒ Cape Toe
   - Ⓓ Princess Bay
   - Ⓔ Isthmus of Jog
   - Ⓕ Bay of Gulls
   - Ⓖ Fish Bay
   - Ⓗ Shark Strait
   - Ⓘ Island Airport

2. Find one other strait/channel. Draw four or five X's in it.

Note: For questions 3, 4, 5, and 6, be sure to consult the key to the map.

3. Draw a straight road between Faith and Hope.

4. Show the symbol for an airport at I.

5. Draw a straight railroad line between Hope and Gulf City.

6. Draw a bicycle path from Mountain View to Faith to Lake Port.

7. Lightly color all the waters blue.

READING AND INTERPRETING CROSS-SECTION AND AERIAL DIAGRAMS

Background

The visualization of geographical terms and symbols is an important geographical skill. But, it is difficult to visualize what one has never seen. This activity will help students visualize terms, and it will also help them gain experience in interpreting cross-section and aerial view diagrams.

Suggestions for Teaching

1. Photocopy and distribute the glossary at the bottom of this page.

2. Explain and discuss the items in the glossary.

3. Photocopy and distribute the facing page. Point out that the top portion is a *cross-section*. To help students understand cross-sections, use the example of a slice of layer cake. If we look at the cake from the side, we are looking at a cross-section. A simple blackboard sketch would be helpful.

Also point out that the bottom portion of the page is an aerial view. It is a view that would be seen from an airplane. Tell them that many maps are made from photographs taken above the earth.

4. Explain that the numerals at the end of each glossary entry tell on which diagram the feature may be found. The cross-section employs numerals, and the bottom aerial view employs letters. *River*, number 14, has been done to help students get started.

Note: The completed diagrams may be found in the Answer Key.

GLOSSARY OF LAND, WATER, AND SKY TERMS

Beach: The place where the water meets the land (2)

Canyon: A deep cut in the earth's surface through which water often flows (13)

Cliff: Land that rises almost straight up from water or the earth's surface (4)

Cloud: A mass of thousands of tiny water particles that combine into raindrops and fall to earth (6)

Downstream: The direction in which a river flows (C)

Dunes: Hills or ridges of sand found on beaches and deserts (3)

Lake: A body of water surrounded by land (D)

Left Bank: The bank of a river or stream that is on one's *left* when facing downstream (H)

Marsh: Soft, wet land with grass-like vegetation (E)

Mesa: A flat-topped hill with steep sides (15)

Mountain Range: A series of mountains (8)

Pass: A low place through mountains or hills that allows for passage of people, cars, etc. (11)

Peak: The top—usually pointed—of a mountain (12)

Plain: A large and fairly level area of land; generally flat, but may have low hills (17)

Plateau: A large, elevated area of land that is generally flat; may have hills (7)

Pond: A small body of water surrounded by land; usually smaller than a lake (A)

Precipitation: Rain or snow that falls to earth from clouds (5)

Rapids: A place in a stream or river that has obstacles, such as rocks, over which water flows (F)

Right Bank: The bank of a river or stream that is on one's *right* when facing downstream (I)

River: A natural and considerable water flow (14)

Slope: Upward/downward slant on the side of a mountain (18)

Stream: A small, usually narrow, flow of water over the earth's surface (B)

Surf: The water/waves, sometimes rough, that washes over a beach (1)

Swamp: Wet, spongy land with clumps of grass; some parts covered by water; some trees (G)

Trail to Pass: The route one would follow to go through a pass (10)

Valley: A level area of land with upward slopes on each side; has the appearance of a V (9)

Volcano: A cone-shaped hill or mountain with a hollow center from which lava and smoke flow (16)

IDENTIFYING LAND, WATER, AND SKY FEATURES

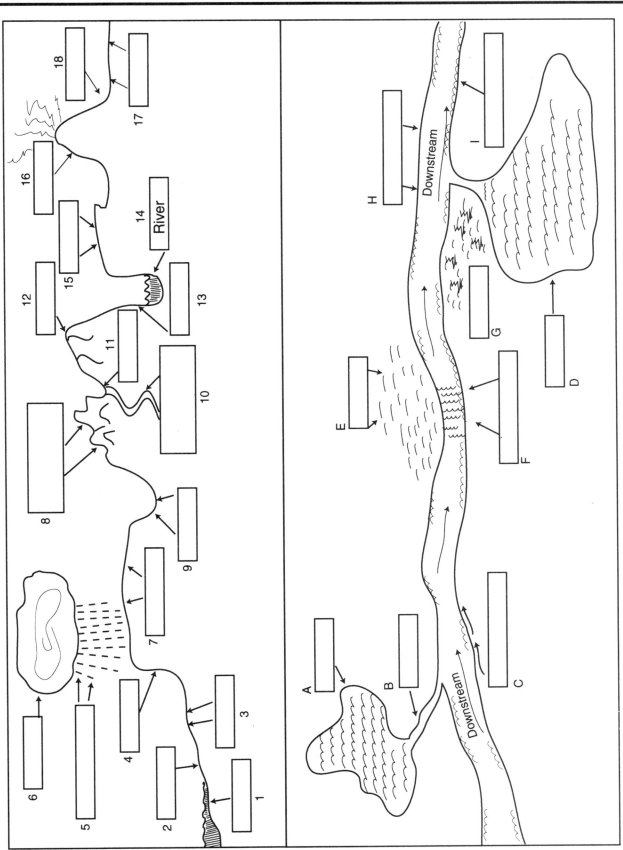

River

Downstream

Downstream

RECOGNIZING BOUNDARIES AND ROUTES OF TRAVEL

Background

Some map symbols represent real things such as cities, roads, lakes, and so on. These are all things that can be seen. Other map symbols represent things that are abstract—that is, not real, as, for example, boundaries or borders.

One could walk a long, long distance along the boundary between the United States and Canada and rarely see a fence, wall, or monument that tells the location of the boundary between the two countries. Occasionally, a sign might be seen that tells that Canada is on one side and the United States is on the other. And, of course, on the borders of the two countries, especially at road crossings, there will be signs and government officials checking vehicles and passengers.

What are some other kinds of boundaries or routes of travel shown by lines on maps?

• Divisions between properties are shown on maps by lines of one kind or another. Sometimes, neighbors will erect fences or walls between their property and a neighbors. This makes the boundary tangible.

• On maps, latitude and longitude are shown by lines. Special lines are used to show the location of various circles including the Equator, the Tropic of Capricorn, the Arctic Circle, and other lines such as the International Date Line. All of these lines are abstractions.

• Weather maps use lines to show otherwise invisible weather fronts and temperature zones.

• Airplane routes and ocean routes are also shown by lines of one kind or another.

• Compass routes may also be shown on maps. There are no roads or trails to follow. Compass routes are completely abstract.

Determining Distance on a Map-Distance Table

Students should learn how to use the distance table found on most state and regional maps. The procedure for using the table is as follows:

1. Find in the alphabetical listing the name of the community from which you wish to start a trip.

2. Also find in the alphabetical listing the name of the community to which you want to travel.

3. With a light pencil, draw a line from one of these communities across the column.

4. Then, with a light pencil, draw a line from the other community down the column.

5. The road distance between the two communities is listed where the two lines intersect.

6. The example on the road distance chart shows that the road distance between Hackensack and Cape May is 159 miles.

7. Notice also that the table lists the index square in which each community can be found on the map. For example, Hackensack is located in index square E-12 on the map.

MAP SYMBOLS FOR THINGS THAT CANNOT BE SEEN

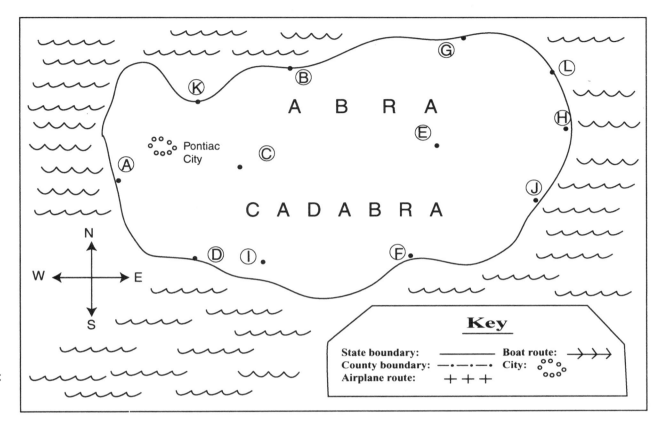

In the previous two lessons you learned how to recognize and interpret symbols for real things such as towns, peninsulas, and railroads. In this activity you will learn how to recognize things you can't see, such as boundaries and ocean and airplane routes.

To Do:

1. Lines between states are usually shown by unbroken lines. Draw a state line from A to H. What two states are separated? _____ and _____

2. Within a state there are counties. Counties may be shown by broken lines. See the key to the map.

a. Draw county lines in Abra from C to B and from E to G.

b. The names of the counties of Abra from left to right (west to east) are PAMELA, BEVERLY, and LORNA. Print these names carefully in large print in the counties.

3. Draw county lines in Cadabra from D to C and F to E.

4. The names of the counties of Cadabra from left to right are ROBERT, THOMAS, and ANTHONY. Print these names in large print.

5. The boundaries of cities may also be shown by lines or symbols. You can see the symbol for Pontiac City in PAMELA. The symbol of the city is like that in the map's key.

In each county draw a symbol for a city, and give each city a short name. Because cities are smaller than counties, their names are printed in smaller print. Print as carefully as you can, and be sure that you follow the example shown in the key.

6. There is an airport in THOMAS at I, and there is another airport in ANTHONY at J. Study the key for the kind of line that would show an airplane route between the two places. Draw the route between I and J.

7. A cruise boat takes passengers from K along the north coast to L. Study the key, and draw the cruise boat's route from K to L. Keep the line close to the coast.

HELP IN TEACHING DIRECTIONS

Background

Following is a helpful way to "drill" directions with students. Use a magnetic compass to determine which wall of the room is closest to being north. Then, post direction signs around the room in large letters: NORTH, SOUTH, EAST, WEST, NORTH-EAST, and so on. Refer to room locations whenever possible. For example, "Tom, please go to the northeast part of the room and bring me the C volume of the encyclopedia."

Another technique is to play "Simon Says" during a break. Have the class stand and face north. Give such commands as, "Simon says raise your east hand," or "Simon says face west." Anyone who doesn't follow the command correctly has to sit down. Students don't like to sit down when everybody else is standing up. Soon students become very quick and accurate in responding to Simon's directions.

Correct any expressions such as "Up north" or "Down south." *Up* is the direction from the center of the earth to the atmosphere. *Down* is the direction from the atmosphere to the center of the earth. Expressions such as, "We went up to Canada," should be corrected immediately. The correct statement would be, "We went north to Canada." Tell your students to be alert to such mis-expressions as they listen to weather broadcasts. It is not uncommon for weather forecasters to say such things as, "As we go down south, we see that temperatures are increasing."

The 16 Basic Directions

The four basic directions are *north, south, east,* and *west*. The directions midway between each of these are *northeast, southwest, southeast* and

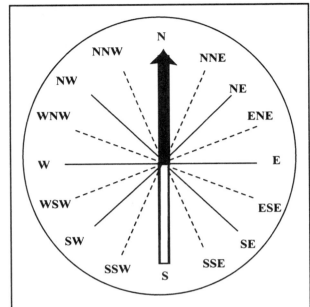

THE 16 POINTS OF THE COMPASS

There are 22½° between compass points. NNE means north-north-east; ENE is east-north-east, and so on.

northwest. Finally, for more accurate directions, we have eight subdivisions: *North-north-east* (NNE), *East-north-east* (ENE), *East-south-east* (ESE), *South-south-east* (SSE), *South-south-west* (SSW), *West-south-west* (WSW), *West-north-west* (WNW), and *North-north-west* (NNW). The diagram above shows the arrangement.

Students may find it helpful to know that a compass needle points toward *magnetic north*, which is located more than 1,000 miles from *polar north* or *true north*. Therefore, when using a compass in the field or when plotting an airplane or sea route, one must take into account the angle of deviation between true north and magnetic north. Later in this book there will be an activity relative to magnetic and true north.

USING DIRECTIONS TO LOCATE CITIES

It is important to know directions when working with maps. This activity will help you to become more skillful in the use of the eight basic directions.

To Do:

The map shows a place that is square-shaped such as the states of Wyoming and Colorado. You are to imagine that you are in the direction circle in the center of the map. In what direction is the city of Holt, shown by the dot below its name? The answer is north (N). So, you should check the N column in the table. Tell the direction each listed city is from the center of the direction circle.

Direction Abbreviations
You should also know the abbreviations of directions used when working with maps. Here is a list.

North: N   Northeast: NE
South: S   Southeast: SE
East: E   Southwest: SW
West: W   Northwest: NW

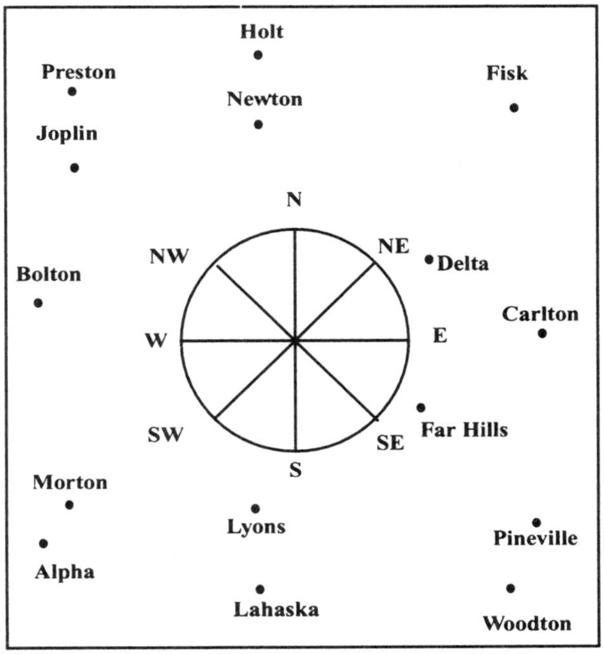

CITY	DIRECTION FROM THE CENTER OF THE CIRCLE							
	N	S	E	W	NE	SE	SW	NW
Preston								
Joplin								
Bolton								
Lahaska								
Carlton								
Far Hills								
Pineville								
Lyons								
Delta								
Morton								
Newton								
Holt	√							
Woodton								
Alpha								
Fisk								

Name: _____ Date: _____

COMPLETING A DIRECTION MAZE

To Do:

 The puzzle below is a direction *maze*. Begin where it says START at the top of the page. Follow the arrows. On every line print the *direction you are walking*. Use abbreviations for the directions as follows:

N: North	E: East	NE: Northeast	NW: Northwest
S: South	W: West	SE: Southeast	SW: Southwest

If you are completely correct when you complete the puzzle, you are AMAZING!!

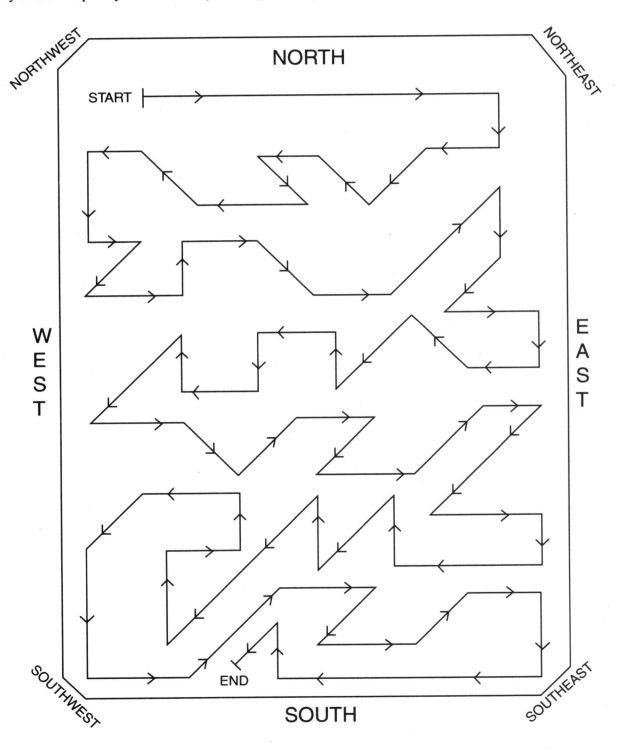

FINDING YOUR WAY AROUND A TOWN

The map shows:

- DUCK LAKE
- WINDSOR STREET
- OFFICES | POLICE STATION | TOWN HALL
- YMCA | CHURCH | LIBRARY
- BROAD STREET
- DRUG STORE | DRESS SHOP | SUPER MARKET
- VARIETY STORE | BAKERY | SHOE STORE
- X Entrance
- MOVIE THEATRE
- WASHINGTON DRIVE
- HOMES
- BALLFIELD
- ONE WAY
- PLEASANT ROAD
- HOSPITAL | BANK | POST OFFICE
- MAIN STREET
- SWIM POOL | SKATING RINK | PLAY GROUND
- TOWN PARK
- STREET — HOMES
- DRIVE — Entrance
- HOTEL | RESTAURANT | FIRE HOUSE
- 1 2 3 4 5 6 7 8 GARDEN APARTMENTS 9 10 11 12 13 14 15 16
- XX
- OAK DRIVE
- ONE WAY
- HIGH SCHOOL
- ELEMENTARY SCHOOL
- SCALE 0 50 100 feet
- N W E S

To Do:

1. In which *direction* would you walk to go directly from the Elementary School to the

a. Bank? _____   b. Playground? _____

2. In which *direction* could you walk to go from the Town Hall to

a. the Movie Theatre? First, south to Washington Drive, then _____ .

b. Garden Apartment 8? First, south to Pleasant Road then _____ .

3. What place is directly northeast of the Library?

_____

4. Start a walk at X on the west end of Washington Drive.
a. Walk east to the end of the block.
b. Turn north and walk one block.
c. Turn east and walk one block.
d. Turn south and walk two blocks.
e. Turn east and walk to the end of the road. What building is to the north of you?

_____

5. Try this walk. Start at XX on Oak Drive.
a. Walk south for two blocks.
b. Turn west and walk to Windsor Street.
c. Turn north and walk one block.
d. Turn east and walk one block.
e. Turn north and walk one block.

f. Turn west and walk to the third store. What store is at that location?

_____

6. Mr. Jones was driving his car west on Broad Street. When he reached Windsor Street he turned south. A police officer stopped him and gave him a ticket. Carefully examine the map and tell why Mr. Jones was ticketed.

_____

7. Imagine that you went to the Library. After you returned your books, you decided to go to the restaurant on Washington Street. Tell what streets and directions you would take to get there. Use numbers 4 and 5 on this page as examples.

a. _____

_____

b. _____

_____

8. Explain how to walk from the Playground to the Movie Theatre in such a way as to make only one turn.

a. _____

_____

b. _____

_____

DETERMINING DISTANCES WITH A SCALE OF MILES

Background

Distances between points on a map may be determined through the use of the *Scale of Miles* found on most maps. With some instruction, students can become quite adept in the use of this skill.

Following are four steps that students may follow to establish distances on maps that contain a scale of miles.

1. Determine the starting and ending points of the distance to be measured. A starting point may be a town, a crossroad, a bridge, and so on. Students should realize that in the case of a city, the starting point is the symbol that locates the city, not the letters in the city's name. Dots may take several forms depending on the size of the city: for example, ●, ◉, ○, ◎.

2. Lay a piece of paper on the map so that the corner of the paper is on the starting point and the edge of the paper touches the ending point. Make a mark on the paper edge where it touches the ending point.

3. Lay the paper's same edge along the scale of miles so that the corner of the paper is at the beginning of the scale.

4. Read the distance on the scale at the point where the edge of the paper was marked. It may be necessary to do some estimating if the pencil mark does not correspond exactly with a mark on the scale.

The illustration that follows shows the steps outlined above. It would be helpful to show it to the class via a transparency.

Note: A ruler may be used to measure map distances. However, this approach requires relatively sophisticated arithmetic. For example, the scale may show that one inch equals 15 miles. If the points in question are 7/8" apart, what is the distance between them in miles? Multiply 7/8 × 15 for a distance of 13 1/8 miles.

Note: The activity on the facing page utilizes an extra long scale of miles. Such a scale will help students determine any distance on the map without successive markings on the paper edge. Other activities in this book will utilize typical scales.

How to Use a Scale of Miles

The Scale of Miles found on most maps can help you determine the number of miles between two places. To use the scale, follow the steps listed below.

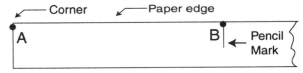

Figure 1

1st: Lay the edge of a blank piece of paper on a map so that the edge cuts through the dots that show where the two places are located. Be sure that the corner of the paper is on one of the marks. (Figure 1 shows dots at A and B.)

2nd: Make a mark on the edge of the paper where it touches B.

3rd: Pick up the paper and lay the same edge along the Scale of Miles. (Figure 2)

4th: At B read the number of miles as shown on the scale. You can see that it is 3 miles from A to B.

Note: If the mark on the paper's edge had been made midway between 2 miles and 3 miles, the distance would have been 2½ miles.

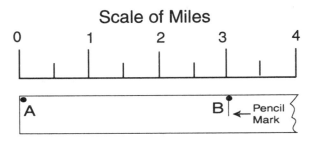

Figure 2

USING A SCALE OF MILES TO FIND DISTANCES ON MAPS

Find the distances between cities on the map. Here are the steps to follow.

1.   Lay a blank piece of paper on the map so that the edge cuts through the dots that show the locations of two cities. Be sure that the corner of the paper is on one of the dots.

2.   Make a mark on the edge of the paper where it touches the second city.

3.   Pick up the paper and lay the edge along the scale of miles. The corner of the paper should be at zero on the scale.

4.   Read the number of miles at the pencil mark. Realize that each inch of the scale is divided into halves. Each half is equal to 8 miles. The example below shows a distance of 40 miles.

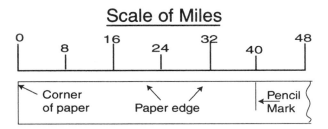

Scale of Miles

To Do:

In each question circle the number that is closest to the distance between the two cities. Do not be concerned if the miles number you select is not the exact mileage.

1.   What distance is the most accurate between

a.   Ivan and Afton?   12   16   24 miles

b.   Afton and Calico? 48   56   64 miles

c.   Drew and Fawn?   48   56   64 miles

d.   Edsel and Booth?   8   16   24 miles

e.   Afton and Booth? 16   24   32 miles

f.   Calico and Hunt? 64   70   80 miles

g.   Afton and Drew?   48   56   70 miles

2.   What is the total distance from Buzz to Port to Booth? _____ miles

3.   What is the total distance from Buzz to Humm to Purr to Buzz? _____ miles

4.   What is the total distance from Gant to Call to Tuff to Gump? _____ miles

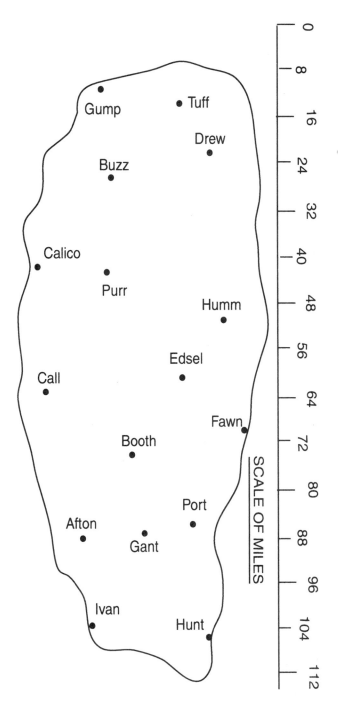

MAP OF ECALPEMOS

READING ROAD MAPS

Background

It is a rare automobile driver who doesn't have at least one road map in the glove compartment of the vehicle. This is because road maps contain a great amount of information. It wouldn't be an exaggeration to say a road map of a state contains hundreds of facts. A partial list of the kinds of information road maps offer follows:

- Roads and their types (number of lanes, kind of surface, county, state, interstate, etc.)

- Locations of communities by index

- Mileage charts telling road distance between cities

- Town and city population parameters

- College locations

- State and national park locations

- Airport locations

- State and county boundaries

- Rest area locations

- Service area locations

- Toll locations

It would be helpful to student understanding of maps to project the legend of a well-developed road map and to point out and explain various symbols. Point out, for example, how service areas are shown. Or, help students discern what can be learned at various tourist information centers.

Student Involvement

The activity on the facing page is concerned with important map reading skills:

- Using a road map legend

- Computing distances through the use of the "mileage-between-towns" system

- Following a route of travel

- Finding places through the use of miles and direction

The map utilizes the mileage-between-towns-and-cities approach for computing distances. For example, the distance between Aspen and Elmwood is shown by the numeral located on the line between the communities. Thus, if several communities are passed on the way to a destination, one only needs to add the numerals to determine the total distance. Point out that large cities are shown by an outline (⌐┐) and that smaller communities are shown by dots.

Note: Call your students' attention to the fact that FOREST CITY is written in capital letters. On road maps, the names of cities are printed larger than those of towns and villages.

It should be of interest and importance for students to realize that numerals indicating distances between towns and cities are more accurate than distances determined by scale of miles. This is because the numeral system takes curves and hills on the road into consideration. The scale-of-miles system considers only the straight-line distance between locations. In mountainous areas, curves and hills can cause a considerable difference in the number of miles between points. Consider the figure below.

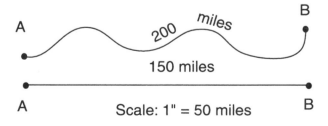

© 2000 by The Center for Applied Research in Education

Name: _____ Date: _____

FINDING YOUR WAY ON A ROAD MAP

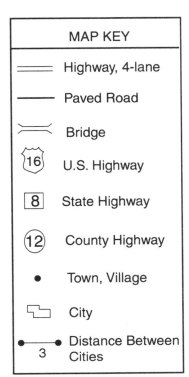

MAP KEY

═══	Highway, 4-lane
───	Paved Road
═╪═	Bridge
(16)	U.S. Highway
[8]	State Highway
(12)	County Highway
●	Town, Village
⌐⌐	City
●—3—●	Distance Between Cities

NOTE:
Distances are computed between "dots."

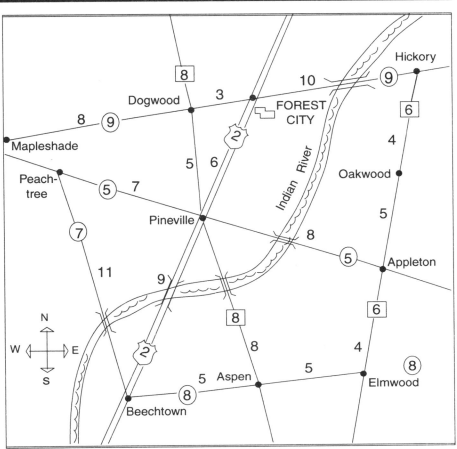

To Do:

1. How many miles is it from

a. Beechtown to Peachtree on County Road (7)?

_____

b. Peachtree to Pineville on County Road (5)?

_____

c. Mapleshade to Hickory on County Road (9)?

_____

d. Elmwood to Hickory on State Road [6]?

_____

e. Beechtown to Pineville on US (2); then Pineville to Appleton on County Road (5)?

_____

2. What five roads cross Indian River?

_____

3. Here is a direction-distance problem:

a. From Elmwood, go north 4 miles on State Road [6].

b. Turn northwest, and go 8 miles on County Road (5).

c. Turn northeast, and go 6 miles on US (2).

Where are you? _____

4. Another distance-direction problem:

a. From Beechtown, go 15 miles northeast on US (2).

b. Turn west, and go 3 miles on County Road (9).

c. Turn south, and go 5 miles on State Road [8].

d. Turn northwest, and go 7 miles on County Road (5).

Where are you? _____

DIRECTION AND DISTANCE AROUND WILDWOOD LAKE

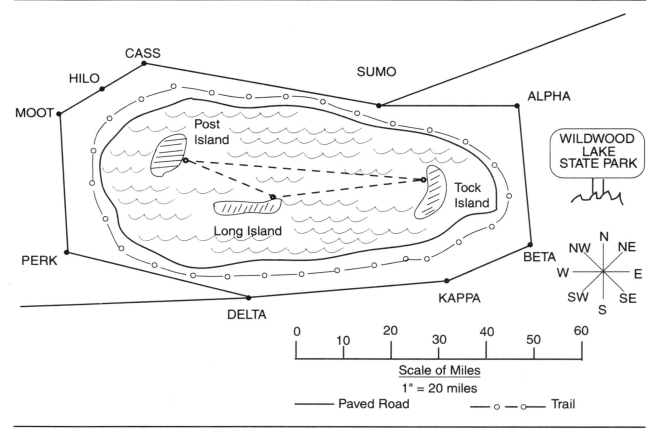

To Do:

1. How many road miles apart are

a. Perk and Delta? _____ miles

b. Alpha and Beta? _____ miles

c. Sumo and Cass? _____ miles

d. Moot and Perk? _____ miles

e. Perk and Beta? _____ miles

2. What direction from Delta is

a. Alpha? _____

b. Moot? _____

c. Kappa? _____

3. What direction from Alpha is

a. Perk? _____

b. Beta? _____

c. Hilo? _____

4. What is the total distance in road miles from Beta to all the towns around the lake and back to Beta?

_____ miles

5. Imagine that you had to post the direction and distance road signs shown below. In what towns would the signs be posted?

a. **Beta 20 miles ▶** Town? _____

b. **◀ Cass 50 miles** Town? _____

c. **Hilo10 miles ▶** Town? _____

d. **◀ Perk 40 miles | Kappa 40 miles ▶**

Town? _____

6. How many miles is the round trip from Post Island to Post Island? _____ miles

USING AN INDEX TO LOCATE PLACES

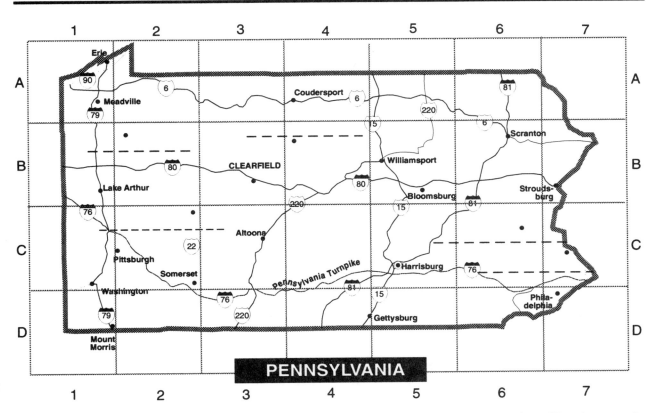

The *Index System* used for locating places on maps is easy to use and very convenient. Here is an explanation telling you how to use it:

a.   Notice the letters A, B, C, and D on the sides of the map. Also notice that across the bottom and top of the map are the numerals 1, 2, 3, 4, 5, 6, and 7.

b.   The letters and numbers give names to the squares, so the first square at the top left of the map is A1. Notice that Meadville is located in that square. The square at the bottom left is D1. The town of Mount Morris is in that square.

To Do:

1. In what square is each of the listed communities? The first one has been done to help you get started.

Community	Square	Community	Square
Mount Morris	D1	**Clearfield**	
Altoona		**Coudersport**	
Williamsport		**Gettysburg**	
Bloomsburg		**Philadelphia**	
Stroudsburg		**Scranton**	
Harrisburg		**Somerset**	

2. Various places—shown by dots—and the squares in which they are located are listed below. Label each place on the map where dashed lines have been drawn.

a.   Oil City: B2   b.   Punxsutawney: C2

c.   Emporium: B4   d.   Allentown: C6

e.   New Hope: C7

3. Through what index squares does Highway 220 pass?

_____ _____ _____ _____

4. Through what index squares does Highway 6 pass?

_____ _____ _____ _____

_____ _____ _____

DEVELOPING A SENSE OF PLACE

2-1   **Place-Name Geography (Instructor)** 22

2-2   **United States Map** 23

2-3   **Crossword Puzzle of the United States I** 24

2-4   **Crossword Puzzle of the United States II** 25

2-5   **All About Rivers (Instructor)** 26

2-6   **The Mississippi Valley** 27

2-7   **Wordsearch on East-of-the-Mississippi-River Capitals** 28

2-8   **The Eastern States and Their Capitals** 29

2-9   **Wordsearch on West-of-the-Mississippi-River Capitals** 30

2-10 **The Western States and Their Capitals** 31

2-11 **The World's Major Land and Water Divisions (Instructor)** 32

2-12 **The World's Continents and Oceans** 33

PLACE-NAME GEOGRAPHY

Background

One area of geography skills and knowledge that doesn't receive the attention it deserves is *locational* geography or, as it is often called, *place-name* geography. There are many well-educated people who could not go to a blank map and point out such places as the Baltic Sea, Bolivia, Ethiopia, and the Suez Canal. The new countries that have separated from the former Soviet Union and those that have formed from the breakup of Yugoslavia and Czechoslovakia are virtually unknown to many high school and college graduates.

In a world that is, figuratively, growing smaller, knowledge of locations is crucial to understanding world and domestic problems. Can one really understand the 1999 problems in Serbia, Kosovo, and Albania without knowing where these countries are and the geographical and historical factors that have contributed to a variety of ethnic and religious groups settling there with little integration?

How many high school graduates could go to a blank map of the United States and name the states? How many could tell the approximate size, population, and major physical features of their own state?

Suggestions for Teaching

1. In teaching place names, some basic questions should be emphasized. Where is the place located? What are its geographical features? What is the geographical context in which the place is located? How is the place important historically and contemporarily because of its location?

2. Establish a list of place names to be learned. Search the social studies books that will be used. History books and, of course, geography books, are loaded with names. For example, if the class will be studying World War II, determine which places are important to that topic and should become part of a student's knowledge base. Places that come to mind

are the United Kingdom of England, Scotland, Wales, and Northern Ireland; the English Channel; the Rhine River; the Netherlands; and so on. Divide the list according to topics.

3. Photocopy and distribute the list. Give a brief overview of the list with reference to maps.

4. As the students encounter the names during the day-to-day studies, offer a brief background on each of them. For example, in speaking of the English Channel point out that it has long been one of Great Britain's important defenses, that it varies from 20 to 100 miles in width, and that there is now a tunnel connecting England with France.

5. Make blank wall maps: no names, but dot symbols for cities, symbols for permanent physical features, and other necessary symbols.

6. Each day take 5 to 10 minutes to drill on the listed names. Here is an example: Point to the Adriatic Sea and say, "Sea?" Call on someone to identify the place. Not always, but frequently ask for a fact. Then, point out another place. Keep the drill brisk.

7. After a reasonable length of time, perhaps a week, administer a blank-map test.

8. Go on to the next section, and continue as explained above. However, be sure to review past place names to help the students retain previously studied locations.

You can confidently expect that at the end of the school year your students will have *command* of some 100-200 place names, their locations, and their significances.

Student Involvement

The activity on the three following pages will emphasize place names and require students to use direction skills. Also, the combination of place names and directions will help them retain the names and locations of the 50 states.

© 2000 by The Center for Applied Research in Education

UNITED STATES MAP

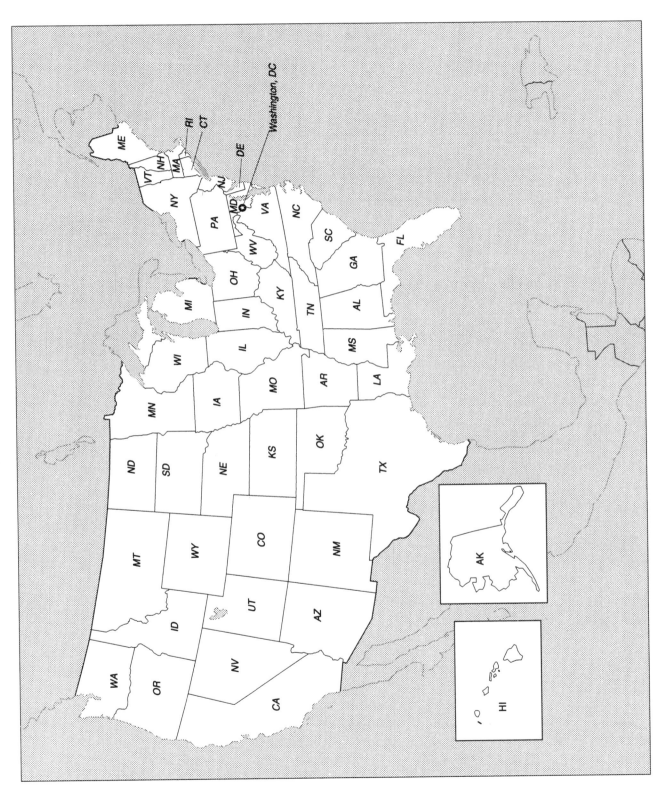

CROSSWORD PUZZLE OF THE UNITED STATES I

ACROSS

1. Pacific Ocean coast; south of Oregon
5. Atlantic Ocean coast; east of New Hampshire
7. East of Washington and Oregon
8. North of Texas (abbr.)
9. Between New York and New Hampshire (abbr.)
10. North of Iowa (abbr.)
12. East of California; south of Utah; west of New Mexico
15. East of Indiana; north of Kentucky
16. Island state in the Pacific Ocean
17. East of New York; south of Vermont and New Hampshire (abbr.)
18. South of Massachusetts; east of New York (abbr.)
19. East of Maryland; southwest of New Jersey (abbr.)
20. South of Washington; west of Idaho
22. South of South Dakota; east of Wyoming; north of Kansas
24. South of Minnesota; north of Missouri (abbr.)
26. South of Montana; east of Idaho
29. Borders the Gulf of Mexico; south of Oklahoma; east of New Mexico
31. Between Alabama and South Carolina
33. North of Oklahoma; south of Nebraska
34. South of Virginia (abbr.)
36. South of Wisconsin; west of Indiana (abbr.)
37. East of Minnesota; north of Illinois
38. Between Nevada and Colorado
40. South of Missouri; north of Louisiana

DOWN

1. South of Wyoming; north of New Mexico
2. East of Texas; west of Mississippi (abbr.)
3. Peninsula state in southeast (abbr.)
4. South of New York; west of New Jersey
5. North of Wyoming; east of Idaho
6. East of California (abbr.)
7. Between Illinois and Ohio (abbr.)
11. North of Indiana and Ohio (abbr.)
13. South of New York; east of Pennsylvania
14. East of Vermont; west of Maine; north of Massachusetts (abbr.)
17. South of Iowa; north of Arkansas (abbr.)
21. Northwest state; north of Oregon (abbr.)
23. South of Massachusetts; east of Connecticut (abbr.)
25. East of Arizona; west of Texas (abbr.)
27. South of Pennsylvania; north of Virginia (abbr.)
28. East of West Virginia; north of North Carolina
30. Between Mississippi and Georgia
32. Arkansas and Louisiana on its west (abbr.)
34. North of Pennsylvania and New Jersey (abbr.)
35. South of North Carolina (abbr.)
37. Ohio on the west; Kentucky on the southwest; Virginia on the east (abbr.)
39. South of Kentucky (abbr.)
40. Extreme northwest of North America (abbr.)
41. North of Tennessee (abbr.)
42. North of South Dakota (abbr.)
43. South of North Dakota (abbr.)

STATES AND THEIR ABBREVIATIONS

Alabama: AL	Illinois: IL	Missouri: MO	Oregon: OR
Alaska: AK	Indiana: IN	Montana: MT	Pennsylvania: PA
Arizona: AZ	Iowa: IA	Nebraska: NE	Rhode Island: RI
Arkansas: AR	Kansas: KS	Nevada: NV	South Carolina: SC
California: CA	Kentucky: KY	New Hampshire: NH	South Dakota: SD
Colorado: CO	Louisiana: LA	New Jersey: NJ	Tennessee: TN
Connecticut: CT	Maine: ME	New Mexico: NM	Texas: TX
Delaware: DE	Maryland: MD	New York: NY	Utah: UT
Florida: FL	Massachusetts: MA	North Carolina: NC	Vermont: VT
Georgia: GA	Michigan: MI	North Dakota: ND	Virginia: VA
Hawaii: HI	Minnesota: MN	Ohio: OH	Washington: WA
Idaho: ID	Mississippi: MS	Oklahoma: OK	West Virginia: WV
			Wisconsin: WI
			Wyoming: WY

CROSSWORD PUZZLE OF THE UNITED STATES II

To Do:

1. Complete the crossword puzzle of the 50 states by using the clues listed under ACROSS and DOWN. For example, 1 ACROSS says, "Pacific Ocean coast; south of Oregon." The map tells you that there are three states on the coast, but only one is south of Oregon. So, California fits exactly into the spaces on the puzzle.

2. If *abbr.*, which means *abbreviation*, is at the end of an entry, study the table and find the abbreviation. Then print the abbreviation in the two spaces allotted for abbreviations. Example: Number 2 DOWN asks for a state that is "East of Texas; west of Mississippi (abbr.)" Louisiana (LA) fits that description. So LA is printed in 2 DOWN. Note that the L is already printed as the third letter in CALIFORNIA.

3. When an entry says *south of* or *north of*, etc., it is referring to a state that is next to and borders the state. For example, in 1 above, California is immediately south of Oregon and is on the Pacific Ocean coast.

25

ALL ABOUT RIVERS

Background

Rivers are very useful natural resources. They supply water for drinking, cooking, washing, swimming, boating, fishing, and power. They are important routes of transportation. This was especially true in colonial times when they were the main "roads" the colonists used. Many, if not most, early settlements were made along rivers or some other body of water such as a bay or gulf.

Students should be aware of the nomenclature of rivers. Following are some of the more frequently used terms.

Source: This is where a river begins. Sometimes the beginning is a spring. Sometimes a source is the place where several streams come together to form a larger stream. Some rivers get their start at lakes.

Mouth: This is where a river ends. The end may be a junction with another larger river. Or, the mouth may be on the coast of a bay, gulf, lake, sea, or an ocean.

Delta: Often, rivers do not empty directly into a large body of water. This is especially true if the land where the river ends is low. Before coming to an end, the river may divide into a number of smaller streams. From the air the streams and the main river may look like an outspread fan. Most rivers carry sediment. When the flow of the river slows down, the sediment drops to the bottom. Over the years the sediment buildup rises above the water and resembles a small island.

So, the fanning out of the river and the soil buildup form a *delta*, named after the fan-shaped Greek letter *D*. The Mississippi River delta on the Gulf of Mexico where the city of New Orleans is located is a classic example of a delta.

Tributary: A smaller stream that joins a larger stream is called a *tributary*. That is, it contributes water and sediment to the larger stream. Some tributaries are large rivers. The Ohio River joining the Mississippi River is an example of a very large river flowing into an even larger river.

Upstream: When one goes upstream, one is going against the current. *Downstream* is going with the current. The slope of the land determines the direction of flow. A river can flow in any direction. Some students find it difficult to understand that rivers can flow north. Why do they have trouble with this concept? Because they mistakenly equate north with going "up" and south with going "down."

The Nile River is an excellent example of a north-flowing river. It flows "downhill" from the highlands of interior Africa to the lowlands along the southern slope of the Mediterranean Sea. Generally, then, cities located upstream have higher elevations than cities located downstream on the same river. The exception to this may occur if a downstream city is on a high bluff overlooking the river as is the case of Quebec on the St. Lawrence River.

Channel: This is the deeper part of a river bed. The water to the left and right of a channel is more shallow than the channel. Ships traveling up and down rivers always use the channel; if they didn't, they would flounder on the other parts of the river bed. It helps to visualize a channel as a kind of "rut" in the river bed. Often channels have to be dredged because sediment builds up and decreases the depth of the channel.

Students should be aware of other uses of the word *channel*. A water passage between two larger masses of land is also called a channel, as, for example, the English Channel between Great Britain and France.

Suggestions for Teaching

1. Most of the large rivers of the United States are shown on the map on the facing page. After explaining, illustrating, and discussing the foregoing, make applications to the rivers shown on the students' photocopied maps.

2. Students should not only understand rivers, but also learn their locations. It would be helpful to include rivers in their "map memory" list. Use blank maps to "drill" students on rivers' locations.

Name: _____ **Date:** _____

THE MISSISSIPPI VALLEY

Rivers of the Mississippi Valley

All About Rivers

- The *source* is where a river begins.

- The *mouth* is where a river ends.

- From the source, the river flows *downstream* toward the mouth. *Upstream* is the opposite of downstream. Upstream is against the current and toward the source.

To Do:

1.a. Print an S at the *source* of the following rivers: Rio Grande, Arkansas, Big Horn, Wabash, Alabama, Cumberland.

b. Draw an arrow (→) pointing *downstream* on all the rivers in 1.a. above.

c. Print an arrow facing *upstream* on the following rivers: Pecos, Brazos, Illinois, North Platte, Des Moines.

d. Print an M at the *mouth* of the following rivers: Sabine, Yellowstone, Pearl, Wisconsin.

2. A small river that joins a larger river is called a *tributary*.

What river is a tributary to the

a. Arkansas River? _____

b. Missouri River? _____

c. Rio Grande River? _____

d. Yellowstone River? _____

3. Print a check (✓) in front of the city that is farthest *upstream* on each river:
a. Arkansas R.:   ___ Tulsa   ___ Wichita

b. Missouri R.:   ___ Bismarck   ___ Pierre

c. Trinity R.:   ___ Fort Worth   ___ Dallas

27

Name: _____ Date: _____

WORDSEARCH ON EAST-OF-THE-MISSISSIPPI-RIVER CAPITALS

A	U	G	U	S	T	A	B	C	O	N	C	O	R	D
T	H	K	F	T	A	M	B	O	S	T	O	N	U	M
L	A	Y	R	N	L	A	D	L	G	J	L	M	X	A
A	R	W	A	I	L	L	P	U	U	M	V	D	R	D
N	T	H	N	Y	A	B	P	M	J	S	R	Y	R	I
T	F	M	K	V	H	A	B	B	M	D	E	T	I	S
A	O	P	F	L	A	N	S	I	N	G	Z	W	C	O
Q	R	P	O	M	S	Y	R	A	L	E	I	G	H	N
S	D	Y	R	F	S	K	F	E	M	O	Y	P	M	Z
M	O	N	T	P	E	L	I	E	R	F	G	I	O	H
O	V	W	E	R	E	T	R	E	N	T	O	N	N	A
N	E	M	K	L	N	U	I	R	D	E	B	A	D	R
T	R	T	D	I	J	A	C	K	S	O	N	P	F	R
G	H	C	H	A	R	L	E	S	T	O	N	L	F	I
O	B	H	F	X	N	A	S	H	V	I	L	L	E	S
M	A	N	N	A	P	O	L	I	S	B	N	M	J	B
E	Y	P	R	O	V	I	D	E	N	C	E	A	H	U
R	C	O	L	U	M	B	U	S	D	F	H	B	T	R
Y	D	S	P	R	I	N	G	F	I	E	L	D	Q	G
W	I	N	D	I	A	N	A	P	O	L	I	S	K	H

To Do:

1. The WORDSEARCH puzzle above contains the names of 26 state capitals east of the Mississippi River. The names of the state capitals are on the map on the opposite page.

2. Circle the names of the capitals. Some are printed *across*, and some are printed *down*. One capital, *Atlanta*, has already been circled to help you get started.

3. After you have solved the puzzle, complete the table of states and capitals on the opposite page. Print the abbreviations of the states' names in the columns labeled *State*.

Name: _____ Date: _____

THE EASTERN STATES AND THEIR CAPITALS

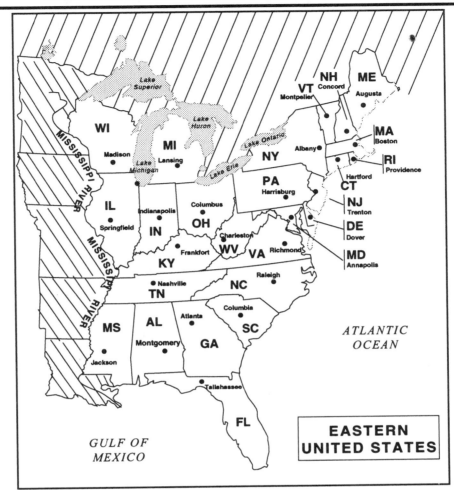

To Do:

Study the map to find the state that has one of the capitals listed below. Write the abbreviation of the state's name on the blank line opposite the name of the capital.

Capital	State (abbr.)	Capital	State (abbr.)	Capital	State (abbr.)
Madison		Augusta		Lansing	
Albany		Montpelier		Concord	
Atlanta		Boston		Tallahassee	
Montgomery		Raleigh		Jackson	
Nashville		Frankfort		Indianapolis	
Charleston		Columbia		Annapolis	
Harrisburg		Hartford		Columbus	
Richmond		Springfield		Providence	
Trenton				Dover	

29

Name: _____ Date: _____

WORDSEARCH ON WEST-OF-THE-MISSISSIPPI-RIVER CAPITALS

B	C	F	X	Q	H	O	N	O	L	U	L	U	W	Y	Z
O	L	Y	M	P	I	A	K	N	Z	S	B	S	D	C	H
K	C	M	E	H	E	W	B	D	M	A	I	A	E	A	G
L	S	P	X	O	Y	T	F	E	H	L	S	N	S	R	L
A	T	C	H	E	Y	E	N	N	E	E	M	T	█	S	V
H	█	B	M	N	H	V	M	V	L	M	A	A	M	O	I
O	P	O	P	I	E	R	R	E	E	G	R	█	O	N	K
M	A	I	C	X	S	Y	R	R	N	Q	C	F	I	█	Z
A	U	S	T	I	N	O	S	W	A	V	K	E	N	C	T
█	L	E	B	A	T	O	N	█	R	O	U	G	E	I	O
C	S	A	C	R	A	M	E	N	T	O	D	P	S	T	P
I	P	L	I	T	T	L	E	█	R	O	C	K	O	Y	E
T	S	A	L	T	█	L	A	K	E	█	C	I	T	Y	K
Y	J	E	F	F	E	R	S	O	N	█	C	I	T	Y	A
J	U	N	E	A	U	F	K	L	I	N	C	O	L	N	S

To Do:

1. The WORDSEARCH puzzle above contains the names of 22 state capitals west of the Mississippi River, as well as St. Paul and Baton Rouge, which are east of the river. The names of the capitals are on the map and table on the opposite page.

2. Circle the names of the capitals. Some are printed *across*, and some are printed *down*.

3. Nine of the capitals have two or three words in their names. In the puzzle, a block (■) separates the words. Be sure to circle the entire name. Salt Lake City has been circled to help you get started.

4. After you have solved the puzzle, complete the table of states and their capitals that is on the opposite page. Print the abbreviations of the states' names in the columns labeled *State*.

THE WESTERN STATES AND THEIR CAPITALS

To Do:

Study the map to find the state that has one of the capitals listed below. Write the abbreviation of the state's name on the blank line opposite the name of its capital.

Capital	State (abbr.)	Capital	State (abbr.)	Capital	State (abbr.)
Olympia		Denver		Des Moines	
Jefferson City		Sante Fe		Salem	
Sacramento		Bismarck		Little Rock	
Pierre		Boise		Baton Rouge	
Carson City		Lincoln		Honolulu	
Topeka		Helena		Juneau	
Cheyenne		Oklahoma City		Austin	
Salt Lake City		Phoenix		St. Paul	

THE WORLD'S MAJOR LAND AND WATER DIVISIONS

Background

Five of the seven world's continents can be easily recognized; they have distinct shapes and locations. Australia is an island. North America and South America are all but separate; only the narrow Isthmus of Panama connects them. Africa is basically an island, except for the narrow Isthmus of Suez that connects Africa to Asia. Some would argue that Africa is a total island because the Suez Canal is a waterway. There is no question that Antarctica is an island continent. It is completely surrounded by water and is a considerable distance from other continents.

Europe and Asia are one great land mass known as Eurasia. But, a careful look at the map shows that Europe is a massive peninsula made up of smaller peninsulas. The Ural Mountains, the Black Sea, and the Caucasus Mountains serve as natural boundaries between Asia and Europe. Historically and culturally those natural barriers have made the two areas distinctly different. Despite the relatively new inventions of airplanes, railroads, and ships, Europe and Asia are still two distinct places.

Suggestions for Teaching

1. Display a simplified globe that clearly shows the continents. Point out each continent, and give brief descriptions of their locations.

2. Display a flat map of the world that shows all the continents. Ask students what differences they notice between the globe representations and the flat map presentations. Explain the distortions that occur when a round world is shown on a flat surface.

3. Make a cylinder of the flat map. Show that the two ends (east and west) meet, but not north and south. This results in the equatorial regions being reasonably accurate in size and shape, but the north and south ends being distorted. For example, Antarctica appears to stretch far east and west. If the flat map shows curved lines of longitude, the distortion will not be as great. However, the shapes of the continents will be altered.

4. On a map that does not have curved lines of longitude, distances will be greatly distorted in the north and south.

5. Explain that, for purposes of easy location and climatic relations, maps usually show lines that are numbered. If lines weren't shown, it would be very difficult to describe the location of a particular place. To show the reality of this, make a mark on an otherwise unmarked ball. Turn and twist the ball. Then ask someone to describe the location of the mark. It will be very difficult to do without reference points.

6. Point out that the equator divides the world into two parts called *hemispheres*. So, we know where to look if reference is made to the *Northern Hemisphere* or the *Southern Hemisphere*. But we wouldn't know *where* to look in those hemispheres for a particular place. We need more lines: horizontal and vertical.

Note: Later in the book there will be activities that explain lines of latitude and longitude.

7. Explain that the world has only one great ocean, but that ocean is divided into smaller oceans: Atlantic, Pacific, Indian, and Arctic. Sometimes there is reference to an Antarctic Ocean. However, the reality is that the so-called Antarctic Ocean is really made up of southern extensions of the Atlantic, Pacific, and Indian Oceans.

THE WORLD'S CONTINENTS AND OCEANS

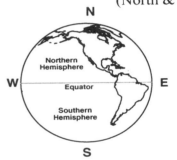

The Hemispheres
(North & South)

To Do:

1.   The world is made up of seven great land masses called *continents*. Their names are given on the map. Name the continents.

_____   _____

_____   _____

_____   _____

_____

2.   Two continents are named on the same land mass. The name of this combination of continents is printed in parentheses ( ). What is its name?

_____

3.   Name the four oceans shown on the map.

_____   _____

_____   _____

4.   The world is divided into two parts. What is the name of the line that divides the parts?

_____

5.   Study the small map of the world. What is the name of the northern part? _____

_____ What is the name of the southern part? _____

6.   Here are some True (T)-False (F) questions. Circle either the T or F after each sentence.

a.   The Southern Hemisphere has more land. T   F

b.   More water is in the Southern Hemisphere.
                                                          T   F

c.   Australia is entirely within the Southern Hemisphere.   T   F

d.   South America and Africa are in both the Northern and Southern Hemispheres.   T   F

e.   The Pacific Ocean separates North America and Africa.   T   F

7.   What ocean would you cross to go from

a.   Africa to Australia? _____

b.   North America to Asia? _____

8.   What two continents are crossed by the equator?

_____

_____

9. What continent is directly north of Africa?

_____

WORLD WATER PASSAGES

3-1   The Panama Canal: Some Basic Information (Instructor) 36

3-2   Sail Through the Panama Canal 37

3-3   The Suez Canal: Some Basic Information (Instructor) 38

3-4   Sail Through the Suez Canal 39

3-5   The Dardanelles, Sea of Marmara, Bosporus: Some Basic Information
      (Instructor) 40

3-6   Sail From the Mediterranean Sea to the Black Sea 41

3-7   The Strait of Gibraltar: Some Basic Information (Instructor) 42

3-8   Sail Through the Strait of Gibraltar 43

3-9   The Strait of Magellan: Some Basic Information (Instructor) 44

3-10 Sail Through the Strait of Magellan 45

3-11 Sail Through the Bering Strait 46

THE PANAMA CANAL: SOME BASIC INFORMATION

Background

The Isthmus of Panama is a relatively narrow stretch of land, about 50 miles, but for more than 400 years it was a natural barrier that greatly inhibited world trade and travel.

Following is some geographical and historical information that will help students better understand the Panama Canal and how it came into being.

❏ The first European to cross the isthmus was the Spanish explorer Vasco Nunez de Balboa. He and his men accomplished that prodigious trek in 1513. Upon emerging on the western side of the isthmus they "discovered" an ocean that Magellan later called the Pacific Ocean.

❏ At the time Balboa made his historical journey, the isthmus was a land of mountains and hills, jungles, wild animals, snakes, swamps, and disease-carrying insects.

❏ Some hardy souls made the trek across the isthmus, including California gold-seekers, but for most people it was an impossible obstacle.

❏ In 1848, Colombia, which controlled the isthmus, gave permission for a group of American businessmen to build a railroad across the land and, thus, connect the Atlantic Ocean and Pacific Ocean by rail. The railroad line was completed by 1855 at a cost of about $8,000,000.

❏ In 1882, a French company attempted to build a canal across the isthmus, but its effort failed.

❏ The new country of Panama, that broke away from Colombia with the help of the United States, gave permission to the United States to build a canal. We were granted control of a ten-mile strip of land that came to be known as the *Canal Zone*.

❏ The building of the canal, which began in 1907, was completed in 1914. The first ship to cross the canal was the cargo-passenger ship *Ancon*.

❏ A problem as significant as the engineering of the canal was the disease carried by rats and insects that ravaged canal workers. Colonel William C. Gorgas led the fight to rid the Canal Zone of disease, especially *yellow fever*, carried by mosquitoes. If the

health problem had not been eliminated, it would have taken many more years to make the "Great Cut," as it has been called.

❏ Before the Panama Canal was built, the trip from New York to San Francisco was made around the southern tip of South America—about 13,000 miles. The canal shortened the journey to about 5,000 miles—a saving of 8,000 miles. This great reduction in miles to sail resulted in the saving of weeks of travel and expense.

❏ The Spanish-American War of 1898 made it even more evident that a canal across the isthmus would be of immense military value to the United States. With a canal, it would be much easier and less time-consuming for United States forces to sail from one ocean to the other. The first war to prove the value of the canal was World War II, fought in Europe and Africa across the Atlantic Ocean and in the far Pacific Ocean.

Information About the Panama Canal Today

❏ 50.72 miles in length; lifts ships 85' above sea level through a series of locks, three on the Atlantic side and three on the Pacific side; each lock is 1,000' long, 100' wide, and 40' deep; the average ship passes through in 15 to 20 hours; more than 10,000 ships make the crossing each year. *Note*: In 1999, according to a treaty (1978) between Panama and the United States, Panama gained full control of the canal and American troops were withdrawn. However, the United States will defend Panama and the canal against any attacks from foreign nations.

Student Involvement

Use the basic information on this page to orient students to the historical and geographical background of the Panama Canal. It would be helpful to have a wall map to show various destinations and routes-of-travel to the canal. For example, from New Orleans to Callao, Peru, via the Strait of Magellan is 11,313 miles. Via the Panama Canal, the distance is 3,200 miles—a saving of 8,113 miles.

Note: Callao is located on the student map.

SAIL THROUGH THE PANAMA CANAL

A _____

1 _____

C _____

E _____

3 _____

San Francisco

New Orleans

New York

Lisbon

B _____

Sydney

4 _____

Panama
Canal

Callao

D _____

F _____

2 _____

THE PANAMA CANAL

Strait of Magellan

To Do:

The building of the Panama Canal was one of the greatest engineering accomplishments in history. Although we had cross-country railroads that could take travelers and freight from our east coast to our west coast, there was no easy way for ships to travel, for example, from New York to San Francisco. To do that, ships sailed around the southern tip of South America—a very long sail in terms of time and distance.

1.   Complete the map by labeling the following:

A: Asia   E: Europe

B: Australia   F: Africa

C: North America   1 and 2: Atlantic Ocean

D: South America   3 and 4: Pacific Ocean

2.a. On the line (- - -) that goes through the Panama Canal write: 5,000 miles.

b.   On the line that goes around the southern tip of South America write: 13,000 miles

c.   Question: How many miles are saved by using the Panama Canal? _____ miles

3.   The diagram below shows a cross-section of the Panama Canal. List, from west (Pacific Ocean) to east (Atlantic Ocean), the places a ship would pass on its way through the canal.

a. _____

b. _____

c. _____

d. _____

e. _____

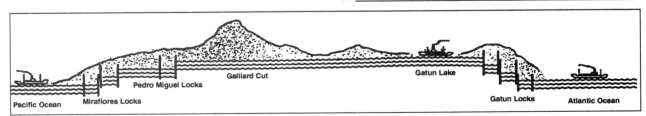

Pacific Ocean

Miraflores Locks

Pedro Miguel Locks

Galliard Cut

Gatun Lake

Gatun Locks

Atlantic Ocean

THE SUEZ CANAL: SOME BASIC INFORMATION

Background

As busy as the Panama Canal is, the Suez Canal is at least twice as busy. It is a major waterway linking Europe, and even the United States, to such places as India and Southeast Asia. Many commodities pass through the canal, but oil may be the most important cargo transported.

The Suez Canal is about twice as long as the Panama Canal. However, it was easier to dig because it is a sea-level canal. No locks are necessary. However, since both banks of the canal are mostly sand, ships are strictly regulated in terms of speed. The waves of moving boats displace the sand that fills the channel. Dredging the canal is a continuing operation, not only because of bank-sand displacement, but also because the wind blows sand from the surrounding desert into the canal.

One of the little known facts about the Isthmus of Suez is that there was a canal connecting the Red Sea and the Nile River more than 1,000 years ago. It can be easily imagined that digging such a canal by hand was an immense undertaking for those times. Neglect, wars, and other problems of upkeep caused the canal to be obliterated. It doesn't take long for wind-blown sand to fill a big ditch.

Following are some additional details relative to the history and geography of the present canal.

❑ Construction of the canal officially began in April 1859. About ten years later, in November 1869, the canal was ready for traffic.

❑ The canal is about 100 miles long and passes through four small lakes. Ships going from north to south enter the canal at Port Said and exit at Suez.

At Suez, ships enter the Gulf of Suez, proceed through the Red Sea, and then into the Arabian Sea.

❑ The canal is capable of accommodating large ships.

- Depth: 196' wide at a depth of 33'
- Width at surface: Average 500'
- Width at bottom: Average 72'
- Mostly single-lane traffic. Ships wait until it is their turn to proceed.
- Average time for passage from end to end: 15 hours. Slow speeds do not produce as much wake as fast speeds; thus, the wear on the sand banks is reduced.

❑ The canal was built for about $92,000,000. However, widening, deepening, and other improvements have added another $322,000,000 for a total cost of $414,000,000.

❑ Control of the canal has changed several times in its history. A Frenchman, Ferdinand de Lesseps, first gained permission from Egypt to construct the canal. Eventually, the canal came under the control of British and French investors. The British guarded the canal during World War I and World War II. In 1956, Egypt took control of the canal. Although Egypt's control was contested by Israel, Great Britain, and France, the canal today remains under the control of Egypt.

❑ An interesting sidelight: In 1980 a tunnel was constructed below the canal. It is an east-west tunnel to be used by automobiles. This makes it possible for drivers of motor vehicles to drive from Africa to the Sinai Peninsula and Israel.

Name: _____ Date: _____

SAIL THROUGH THE SUEZ CANAL

It was in 1869 that a water passage, the Suez Canal, was completed after almost ten years of digging. The new canal connected the Red Sea and the Mediterranean Sea. Before the Suez Canal was built, ships sailing from such countries as France and Norway had to sail south around the southern tip of Africa to reach countries such as India.

To Do:

1.a. Assume that you are sailing from Bombay, India, to New York before the Suez Canal was built. In what direction would you sail before reaching the southern tip of Africa? _____

b. What name is given to the southern tip of Africa?

_____

c. In what direction would you sail from the southern tip of Africa to New York? _____

d. If your ship sailed an average rate of 650 miles per day for the 13,000-mile trip, how many days would the trip take? _____days

2.a. Assume that you are traveling from Bombay to New York through the Suez Canal. What is the name of the passage at the south end of the Red Sea?

_____

b. After you pass through the canal, what sea do you enter? _____

c. What is the name of the passage between the Mediterranean Sea and the Atlantic Ocean?

_____

3. What is the savings in miles of the Bombay-Suez Canal route to New York compared to the Bombay-Cape of Good Hope route? _____ miles

4. If your ship sailed at an average rate of 650 miles a day for the 9,400-mile trip, how many days would the trip take? _____ days. How many days shorter is the Bombay-Suez trip than the Bombay-Cape of Good Hope trip? _____days

39

THE DARDANELLES, SEA OF MARMARA, BOSPORUS: SOME BASIC INFORMATION

Background

• The straits known as the Dardanelles, the Sea of Marmara, and the Bosporus have been important in world history for hundreds of years. They make up the sea route into and out of the countries that border the Black Sea. It follows that the country that controls the passages is in the position of being able to restrict or allow ships to pass through.

• The passages from the Mediterranean Sea to the Black Sea are open all year. They are Russia's entrances into the western world of Europe, Africa, and eastern North America. Much of the foreign policy of Russia—formerly the Soviet Union—has been concerned with control of the passages. Russia has ports on the Arctic Ocean and the Pacific Ocean, but those ports can be ice-locked in winter. The Russians were so concerned about the important waterway that they tried to gain control of the straits after World War II. The United States and other countries came to the aid of Turkey and prevented a Russian takeover.

• The Dardanelles is an interesting strait. The passage is about 37 miles long. Its entrance from the Aegean Sea is narrow—about one mile wide. The narrow width makes the passage easy to defend. Cannon on the shores can easily shell any boat trying to sail through the strait.

• The Bosporus Strait, with a length of about 19 miles, is much shorter than the Dardanelles. Its width ranges from about one-half mile to two miles. A strong current of salt water that comes from the Aegean Sea, through the Dardanelles and the Sea of Marmara, and then through the Bosporus is what makes the Black Sea a salt sea. A bridge crosses the Bosporus into European Turkey.

• The Sea of Marmara is the middle passage of the three passages that connect the Mediterranean Sea to the Black Sea. The sea is about 140 miles long from west-to-east. This distance, when added to the 37 miles of the Dardanelles and the 19 miles of the Bosporus, makes the entire connection between the Aegean Sea and the Black Sea about 196 miles long. The area of the Sea of Marmara is about 4,000 square miles, which is about one-half the size of Lake Ontario in the United States.

• Istanbul, on the western side of the Bosporus, has a population of more than 7,000,000 people; thus, it is one of the world's largest cities. Ankara, the capital of Turkey located on the mainland, has about 3,000,000 people. Istanbul wasn't always so called. For about 1,000 years, it was known as Constantinople, a name derived from the Roman emperor Constantine. Constantinople was built by Constantine and is located on top of the ancient city of Byzantium.

Suggestions for Teaching

1. It would be helpful to use a world map that shows the position of the straits in relation to the world. The background information given above can be used to emphasize the importance of the passages, not only for the countries in their immediate region, but also worldwide.

2. It would be helpful to add the basic place names to the students' repertoire and to drill from time to time as mentioned earlier in this book. The list would include Turkey, Dardanelles, Sea of Marmara, Bosporus, Mediterranean Sea, Aegean Sea, and Black Sea.

Name: _____ Date: _____

SAIL FROM THE MEDITERRANEAN SEA TO THE BLACK SEA

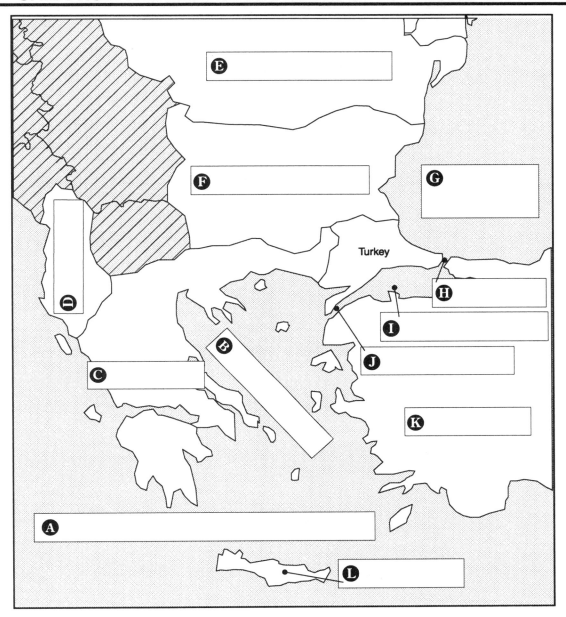

Turkey

The map, when completed, will help you understand the water passageway from the Mediterranean Sea to the Black Sea. Turkey controls the passage. Notice that Turkey is on both sides of the passage.

The passage is important. If there were no passage, the Black Sea would be landlocked.

To Do:

1. Label the following places on the map.

   A Mediterranean Sea   **D** Albania
   B Aegean Sea   **E** Romania
   C Greece   **F** Bulgaria

   G Black Sea   **J** Dardanelles
   H Bosporus   **K** Turkey
   I Sea of Marmara   **L** Crete

2. What two countries shown on the map need the water passages to reach the Mediterranean Sea?

   _____

   _____

3. What two countries shown on the map need the water passages to enter the Black Sea?

   _____

   _____

THE STRAIT OF GIBRALTAR: SOME BASIC INFORMATION

Background

The Strait of Gibraltar was important in World War II and had been important for hundreds of years before that. It is one of the three entrances into the Mediterranean Sea. The early Phoenicians and Romans used the strait in their voyages from the eastern end of the Mediterranean Sea into the Atlantic Ocean. It has been said, and is probably true, that whoever controls Gibraltar controls the Mediterranean.

Following are some helpful facts about the Strait of Gibraltar and the British Royal Colony of Gibraltar.

❑ During World War II, it was the practice of German submariners to sail through the Strait of Gibraltar in quest of Allied shipping. The submarines did not sail on top of the water; they would have been bombarded by British cannon located on top of *The Rock*, as it is called. The submariners cut their motors and floated on a current beneath the surface of the water and into the Mediterranean Sea. Then, on their way out of the Mediterranean Sea, a counter current beneath the surface carried them. By cutting their motors, the German submarines had no sound or vibration and were extremely difficult for the British to detect.

❑ The strait became even more important with the opening of the Suez Canal. Ship traffic into the Mediterranean increased because an all-water route to and from the Middle East and the Far East was now available at the east end of the Mediterranean.

❑ Britain (The United Kingdom of Great Britain and Northern Ireland) is presently in control of The Rock and has had control for about 300 years. About 30,000 people live in the British colony. Most of the people are of Spanish and Italian background. In 1967, the residents had an opportunity to vote on whether to remain a British colony or to join with Spain. The vote was overwhelmingly in favor of staying with the British.

❑ The Rock is more than 1,400 feet above the surface of the sea. A narrow isthmus connects it to the mainland.

❑ Africa is about eight miles across the strait from The Rock. The width of the strait varies from about eight miles to 23 miles.

❑ A little known fact about The Rock is that there is a colony of animals there known as Barbary apes. How the apes happen to be in Europe is a question that never has been fully answered. The rhesus monkey of India is closely related to the Gibraltar apes, and there are also apes in North Africa. It would be interesting for students to conjecture how the apes got to Gibraltar. Among the possibilities is that they were brought there by a passing ship, were released or escaped, and then began to breed.

In any case, the British are very concerned about the welfare of the apes, and they are strictly protected.

Suggestion for Teaching

It would be effective and helpful to student understanding to point out the Rock of Gibraltar on a wall map of the world. The student activity emphasizes a trip south along the west coast of Europe and then eastward into the Mediterranean Sea. However, it can be pointed out that North America, South America, Africa, and Asia all benefit from the Strait of Gibraltar. A chalk line on the map from Boston, for example, through the strait to Athens would help students to see how important the passage is from both a military and trade point-of-view.

SAIL THROUGH THE STRAIT OF GIBRALTAR

To Do:

1. Follow the route of travel that starts in Oslo, Norway, and ends in Rome, Italy. Name in order the places a ship would pass by or through on its way.

Ⓐ _____

Ⓑ _____

Ⓒ _____

Ⓓ _____

Ⓔ _____

Ⓕ _____

Ⓖ _____

Ⓗ _____

Ⓘ _____

Ⓙ _____

Ⓚ _____

Ⓛ _____

Ⓜ _____

_____

Ⓝ _____

Ⓞ _____

Ⓟ _____

_____

Ⓠ _____

Ⓡ _____

Ⓢ _____

_____

The Strait of Gibraltar, shown on the map at Ⓜ, is one of the three water passages into the Mediterranean Sea. The Suez Canal and the Dardanelles are the other two passages.

THE STRAIT OF MAGELLAN: SOME BASIC INFORMATION

Background

In the late 1400s and the early 1500s, Europeans were anxious to find a way to get to the so-called Indies in the Near East and Far East. They wanted the spices, such as cloves, that could produce great profits. However, the routes to the Indies were extremely difficult. The Isthmus of Suez blocked the Mediterranean Sea route. The overland route across Asia was long, rugged, and fraught with dangers. Vasco Da Gama, in 1497-1498, did find a way to reach the eastern lands by sailing south around Africa. However, the voyage was long and arduous.

It was known that the world is round; there had to be a way to sail west across the Atlantic and reach the Indies. But, sea captains/explorers were frustrated in their efforts; they couldn't find a way to get through the great land masses of North America and South America.

Finally, Ferdinand Magellan, a Portuguese in the employ of Spain, determined to sail south and west across the Atlantic Ocean. Perhaps, he thought, there was a southern passage through which they could sail. If this could be done, fame and fortune would follow for the captain and his men.

Facts About Magellan's Voyage

❏   Five ships—the *Santiago*, *Trinidad*, *San Antonio*, *Concepcion*, and *Victoria*—began the voyage from Spain in September, 1519. Only one ship, *Victoria*, completed the voyage.

❏   The crews of the five ships consisted of 240 men. By the end of the voyage, only 18 had survived.

❏   After some four months, the voyagers reached the southern coast of South America. They explored the many inlets and bays in their search of a water passage through the continent. Finally, and with only three ships left, the expedition sailed through the passage now called the Strait of Magellan. They entered a great ocean to which Magellan gave the name Pacific Ocean.

❏   Once in the Pacific Ocean and after a voyage of almost 100 days, Magellan and his crew reached the Mariana Islands. The voyage was continued, and the islands now known at the Philippine Islands came into view.

❏   Magellan became involved in tribal warfare and was killed along with several of his men.

❏   Without Magellan to lead them, the three ships continued on their long voyage. The *Concepcion* was destroyed by fire, and the *Trinidad* was wrecked in a storm. Del Cano, a captain, and the remaining 17 men then sailed the *Victoria* across the Indian Ocean, south along the coast of eastern Africa, around the Cape of Good Hope, north along the west coast of Africa, and, finally, to Spain. The first all-water sail around the world was completed.

Facts About the Strait of Magellan

❏   The passage is at the southern end of South America. The waters are turbulent, and it takes considerable skill to navigate through them.

❏   The passage from the Atlantic to the Pacific is about 350 miles long and varies in width from 2 miles to 20 miles.

❏   South and east of the passage is the island of Tierra del Fuego and several smaller islands. Magellan gave them this name because, as the ships passed through the strait, fires of the natives blazed on the shores and sloping hills. *Note*: *Tierra del Fuego* is Spanish and means *Land of Fire*.

❏   Argentina owns the eastern part of Tierra del Fuego, and Chile owns the western part .

❏   Tierra del Fuego and the many islands near it cover more than 27,000 square miles; the large island is about 18,000 square miles.

❏   Ushuaia is a city on the Argentine half of the island; it is the world's most southern city. Punta Arenas is a Chilean city located on the strait.

❏   Until the Panama Canal was built, the strait or the possible route *around* Cape Horn were the only year-round all-water routes between the east coast of the United States and the west coast.

SAIL THROUGH THE STRAIT OF MAGELLAN

The world map shows the route Ferdinand Magellan took to sail around the world. Magellan was killed in a battle in the Philippine Islands. Only one ship of the five ships that started completed the voyage. Of the 240 men who started the trip in Spain in 1519, only 18 returned.

To Do:

Nine numbered boxes are placed around the map. Write the numbered information that follows in its numbered map box.

1. Five ships leave Spain in 1519.
2. The ships sail to South America.
3. The ships sail the Strait of Magellan.

4. Only three ships are left.
5. Magellan reaches the Mariana Islands.
6. Magellan is killed in battle in the Philippines.
7. *Victoria* sails the Cape of Good Hope.
8. *Victoria* sails north.
9. *Victoria* returns to Spain in 1521.

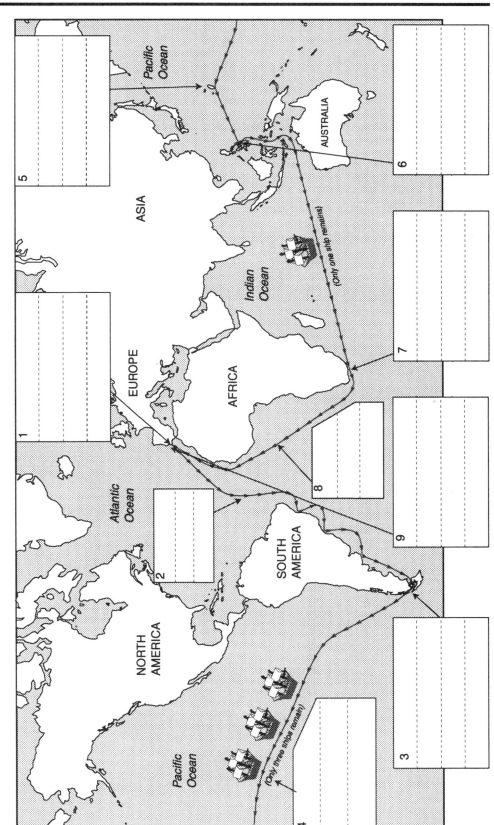

SAIL THROUGH THE BERING STRAIT

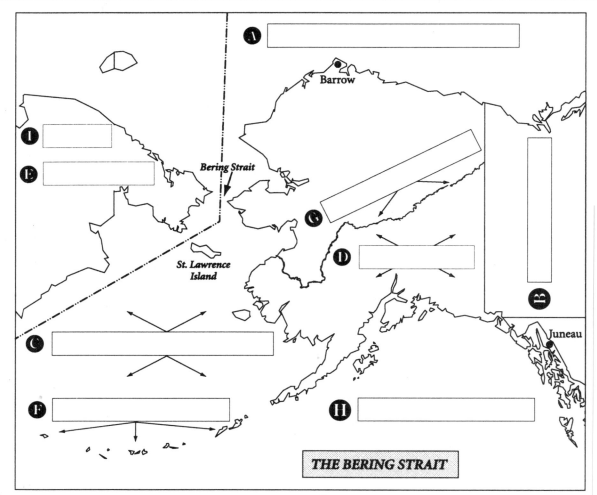

Barrow

Bering Strait

St. Lawrence Island

Juneau

THE BERING STRAIT

© 2000 by The Center for Applied Research in Education

The Bering Strait is a narrow water passage, about 50 miles wide, that connects the Arctic Ocean and the Bering Sea. It also separates Asia and North America. Long ago the strait was a land passage, but eventually the land sank, water came in, and what was an isthmus became a strait.

Historians believe that the first people to come to North America came from Asia when there was a land bridge connecting the two continents. After entering North America some 20,000 years ago, the descendants of the first people gradually spread out. They settled in North America, Central America, and South America.

To Do:

1. The map shows the Bering Strait and the land and water surrounding it. The map will be more useful and understandable after you label it:

A Arctic Ocean
B Canada
C Bering Sea
D Alaska
E Russia
F Aleutian Islands
G Yukon River
H Pacific Ocean
I Asia

2. The — • • — • • — line is the boundary between Alaska and Russia.

To what country does the island of St. Lawrence belong? _____

3. Draw a line (–>–>–>–>) from Barrow at the northern tip of Alaska, through the Bering Strait, east of the Aleutian Islands, to Juneau, capital of Alaska.

4. Color all the water parts of the map a light blue.

© 2000 by The Center for Applied Research in Education

SECTION 4

MAPS THAT SHOW A PATTERN

4-1   **Pattern Maps (Instructor)** 48

4-2   **Traveling by Road and Railroad** 49

4-3   **Using Two Maps to Find Answers to Questions** 50

4-4   **A Pattern Map of the World's Deserts** 51

4-5   **An Altitude Pattern Map of 19 Western States** 52

4-6   **A Pattern Map of Foreign Visitors to the United States** 53

4-7   **The Leading Barley-Producing States** 54

4-8   **Coastlines of the United States** 55

PATTERN MAPS

Background and Student Involvement

Typical maps, as found in textbooks, show a conglomeration of things both human-made and natural—mountains, rivers, roads, cities, and so on. Pattern maps, however, are designed to show a particular distribution of things. Rainfall maps, temperature maps, and agricultural maps are examples of pattern maps. As well as showing a pattern of distribution, some pattern maps also show other elements such as boundaries, water forms, and roads. However, the main focus of a pattern map is not hidden; it is obvious.

Often, two or more pattern maps are shown on the same page. For example, one map may show a distribution of agricultural products and another map may show elevations. It then becomes the task of the reader to determine relationships between the kinds of crops shown and the elevations at which they are grown. Another example: Two pattern maps are displayed. One portrays mineral deposits and the other shows transportation facilities such as roads, railroads, and shipping lines. Selected cities are also shown. Readers can then make inferences related to how the minerals are transported from the mines and quarries, how expensive transportation might be, and so on.

Students should be taught that the first requisite in reading pattern maps is to note the title. What does the map show? Next, the students should examine the key to obtain a general idea of what the map is trying to represent. Next, they should make some quick observations. For example, an elevation map shows clearly that the western United States has more and higher mountains than the eastern part of the country. Another easily apparent observation is that the central part of the country is relatively flat.

Students should realize that pattern maps have their limitations; they may even give false impressions. For example, on a precipitation map there may be a sharp edge between where the 0"-20" rain region ends and where the 20"-40" rain region begins. Of course, there is no sharp difference in rainfall between the two regions. Rather, the two precipitation regions gradually blend. In addition, there may even be pockets of more or less rain within a region, but they are too small to show on a one-page map.

It is helpful for students to make their own pattern maps. The facts that they portray on their maps could be gathered from research or from the instructor. For example, on a blank map of their state that shows counties, they could show the pattern of population distribution. The population could be shown numerically: one county has 40,000 people, another county has 50,000, and so on. Symbols could be used in a more advanced representation. For example, a symbol of a person could represent 10,000 people. Therefore, in the case of the county containing 40,000 people, four symbols would be drawn within the county's borders. It is also good practice for students to take information from a table or graph and then show the information in a pattern map.

The basic steps, then, in the making of a pattern map are as follows:

1.   Obtain a suitable map. (The instructor may have to provide the map.)

2.   Convey the main topic of the map by devising a simple, concise title.

3.   Gather the data for the map. If appropriate, break the data down to manageable segments represented by symbols that have assigned values.

4.   Devise a key to the map.

5.   Arrange the data on the map.

TRAVELING BY ROAD AND RAILROAD

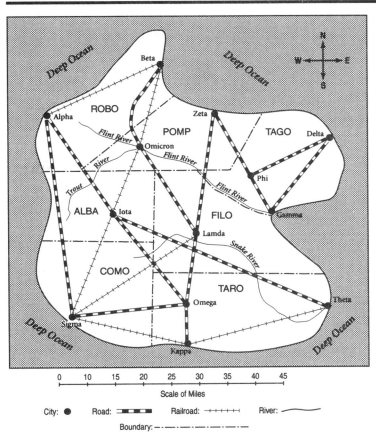

Scale of Miles
0 10 15 20 25 30 35 40 45

City: ●   Road: ▪▭▪▭   Railroad: ┼┼┼┼   River: 〜

Boundary: ▪–▪–▪–

Note: The Flint River is a boundary between two states. If the boundary symbol was put on the river, the river symbol would be covered over.

To Do:

1. What five cities are not on a railroad line?

_____

_____

_____

_____

_____

2. Which one of the following cities probably has the most railroad activity?
__Alpha __Theta __Sigma __Omicron

3. How many times does the road from Iota to Theta

cross the boundaries of FILO? _____

4. What city is located where two rivers join?

_____

5. How many times does the road from Iota to Theta

cross the Snake River? _____

6. Assume that your are traveling only by railroad from Alpha to Theta. Name the cities you would pass through on the route.

_____

_____

_____

_____

_____

7. About how many miles is the road trip from Sigma to Omega?

___25   ___30   ___35

8. Assume that it costs fifteen cents ($.15) a mile for a passenger on the railroad to go from Iota to Sigma. What would be the cost of the trip?

___$3.25   ___$3.75   ___$4.50

9. What two states have the Flint River as a boundary?

_____ and

_____

10. What states share boundaries with the state of POMP?

_____   _____

_____   _____

11. What two cities are both ocean ports and river ports?

_____   _____

12. What is the name of the river that is a tributary to the Flint River?

_____

13. How many bridges across rivers, roads, and railroads would be crossed to go from the Snake River's source to its mouth?

_____

14. If you drove directly from Alpha to Iota to Omega, in what direction would you be driving?

___Northwest   ___Northeast

___Southeast   ___Southwest

USING TWO MAPS TO FIND ANSWERS TO QUESTIONS

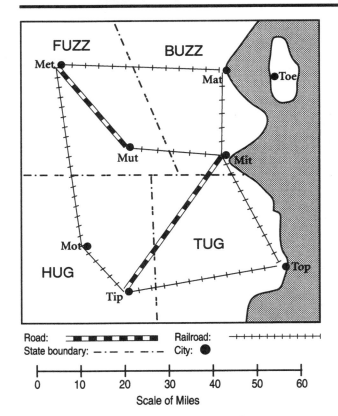

Road: ▰▰▰▰▰   Railroad: ┼┼┼┼┼┼┼┼
State boundary: ─·─ ── ─·─   City: ●

Scale of Miles
0   10   20   30   40   50   60

Coal: ◆   Bauxite: ⊕   Petroleum: ⊠   Copper: ⌓
Forests: ↑   Iron ore: ◎   Mountains: ⋀

Note: Each symbol represents $1,000,000.
Note: Aluminum is extracted from bauxite.

This activity will require the use of the two maps above. For example, the map in the first column shows that M

ut is in the state of FUZZ and that it is located where a road and a railroad meet. However, to know that it is in a forested area you have to study the map in the second column.

To Do:
Circle the letter before the best answer.

1. What kind of workers would you most probably find near the city of Toe?

a. Gold miners   c. Lumberjacks
b. Oil-field workers d. Coal miners

2. Which is the bettter reason why a road or railroad was not built from Mut to Top?

a. The distance is too great.
b. There are too many mountains and forests.

3. What natural resource would the railroad from Tip to Top most likely carry?

a. Coal   c. Zinc
b. Iron   d. Bauxite

4. What is the value of the petroleum in the state of BUZZ? (Study the key to the map.)

a. $4,000,000   c. $6,000,000

b. $5,000,000   d. $7,000,000

5. In the state of HUG, how many millions of dollars more is the bauxite worth than the coal?

a. $1,000,000   c. $3,000,000
b. $2,000,000   d. $4,000,000

6. If it costs $.20 (twenty cents) to carry a ton of ore one mile on a railroad, what would be the cost of shipping 300 tons from Mat to Mit?

a. $800   c. $1500
b. $1200   d. $1600

7. Which two of the following would probably not be shipped from Mat?

a. Bauxite   c. Coal
b. Iron ore   d. Copper

8. A person is looking for work in the petroleum-producing industry. In which state should the person look for the most opportunities?

a. FUZZ   c. HUG
b. BUZZ   d. TUG

9. Which state is most likely to have ski resorts?

a. FUZZ   c. HUG
b. BUZZ   d. TUG

Name: _____ **Date:** _____

A PATTERN MAP OF THE WORLD'S DESERTS

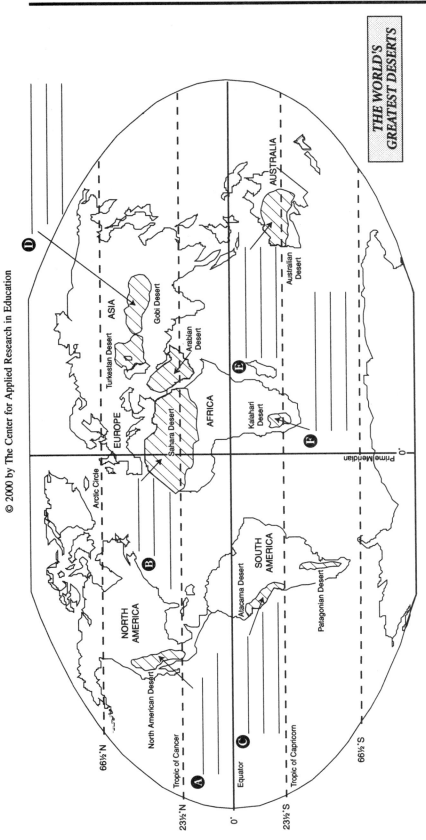

THE WORLD'S GREATEST DESERTS

5. The Atacama Desert is extremely dry: Averages less than 1" of rain per year. Write this fact in C on the map.

6. The Gobi Desert is far from oceans: Very dry, very cold. Write these facts at D on the map.

7. The Australian Desert covers about one-half of the continent: World's second largest desert. Write this fact at E on the map.

8. The Kalahari Desert, about 3,000' above sea level, is the home of African Bushmen: Some grass and small trees. Write these facts at F on the map.

2. Where are most of the deserts?

_____ North of the Equator

_____ South of the Equator

3. The North American Desert contains the continent's lowest point: Death Valley, 282' below sea level. Write this fact in A on the map.

4. The Sahara Desert, which is about the size of the United States, is the largest desert: 3,500,000 square miles. Write this fact in B on the map.

To Do:

1. List each continent's deserts.

CONTINENT	DESERTS
Australia	
Asia	1.
	2.
	3.
Africa	1.
	2.
South America	1.
	2.
North America	

Name: _____ Date: _____

AN ALTITUDE PATTERN MAP OF 19 WESTERN STATES

To Do:

The table tells the highest altitude (elevation, height) in each of 19 western states, including Alaska and Hawaii. You can rearrange the information to make a pattern map.

1. On the lines in each state on the map print the highest altitude for that state. (Do not include the name of the high point.)

Note: Hawaii is so small that you will have to print the information on the line outside the state's outline.

2. How much higher is the altitude of Mount Ranier, Washington, than that of White Butte, in North Dakota? _____ feet

3. What state in the six most eastern states shown has the highest altitude? _____

HIGHEST POINTS IN 19 WESTERN STATES		
State	Highest Point	Name
Alaska (AK)	20,320'	Mt. McKinley
Arizona (AZ)	12,633'	Humphrey's Peak
California (CA)	14,494'	Mt. Whitney
Colorado (CO)	14,433'	Mt. Elbert
Hawaii (HI)	13,796'	Mauna Kea
Idaho (ID)	12,662'	Borah Peak
Kansas (KS)	4,039'	Mt. Sunflower
Montana (MT)	12,799'	Granite Peak
Nebraska (NE)	5,426'	Johnson Township
Nevada (NV)	13,140'	Boundary Peak
New Mexico (NM)	13,161'	Wheeler Peak
North Dakota (ND)	3,506'	White Butte
Oklahoma (OK)	4,973'	Black Mesa
Oregon (OR)	11,239'	Mount Hood
South Dakota (SD)	7,242'	Harney Peak
Texas (TX)	8,749'	Guadalupe Peak
Utah (UT)	13,528'	Kings Peak
Washington (WA)	14,410'	Mt. Ranier
Wyoming (WY)	13,804'	Gannett Peak

HIGHEST ALTITUDE IN EACH OF THE 19 WESTERN STATES

Note: ALT in each state means altitude.

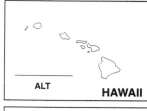

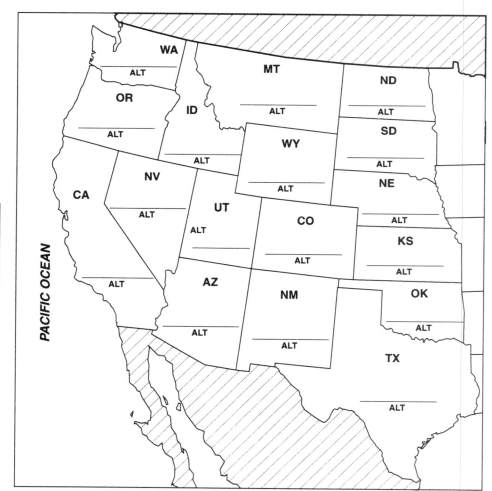

Name: _____   Date: _____

A PATTERN MAP OF FOREIGN VISITORS TO THE UNITED STATES

To Do:

Many foreign people come to the United States for pleasure and business. The pattern map shows some of the cities most visited by foreigners in a recent year.

Take the information from the pattern map and arrange it in the table according to rank. New York had the most visitors, so it is ranked first.

CITY	VISITORS	CITY	VISITORS	CITY	VISITORS
New York City	4,532,000				

THE LEADING BARLEY-PRODUCING STATES

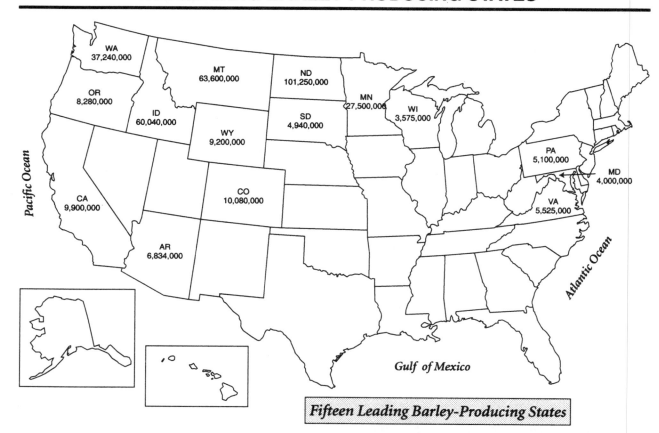

Fifteen Leading Barley-Producing States

The map shows the 15 leading barley-producing states in a recent year. The amount of barley produced in each state is shown by the number within each state. The number represents bushels of barley.

To Do:

1. Scan the map carefully. Then, below and in the next column beside the name of each state, write the number of bushels produced.

2. Which region is the greater barley producer?
___Northeast   ___Northwest

3. What three states are the only major barley producers in the east?

_____

_____

_____

State	Bushels of Barley
North Dakota (ND)	
Montana (MT)	

State	Bushels of Barley
Idaho (ID)	
Washington (WA)	
Minnesota (MN)	
Colorado (CO)	
California (CA)	
Wyoming (WY)	
Oregon (OR)	
Arizona (AZ)	
Virginia (VA)	
Pennsylvania (PA)	
South Dakota (SD)	
Maryland (MD)	
Wisconsin (WI)	

COASTLINES OF THE UNITED STATES

ATLANTIC OCEAN

Maine:
New Hampshire:
Massachusetts:
Rhode Island:
New York:
New Jersey:
Delaware:
Maryland:
Virginia:
North Carolina:
South Carolina:
Georgia:
Florida:
(Atlantic Ocean)
(Gulf of Mexico)
Alabama:
Mississippi:
Louisiana:
Texas:

Washington:
Oregon:
California:

PACIFIC OCEAN

United States
Coastline
Mileage Map

What is a coastline? It is the length of land that borders on an ocean, bay, or gulf. Coastline does not include islands off the coast.

To Do:

1. Take the coastline figures from the table and write them on the blank lines next to or below the names of the states.

2. Florida has two coastlines. What is Florida's combined coastline mileage?

_____ miles

3. Considering all the states except Florida, which state has the longest coastline?

_____ How long? _____

4. What state has the second longest Atlantic coastline? _____

How many miles shorter than Florida's Atlantic coastline? _____

5. What is the total length of the Pacific coast? _____

Atlantic coast? _____ miles; Gulf of Mexico coast? _____ miles

ATLANTIC COAST	LENGTH	GULF COAST	LENGTH
Connecticut	0	Alabama	53
Delaware	28	Florida	770
Florida	580	Louisiana	397
Georgia	100	Mississippi	44
Maine	228	Texas	367
Maryland	31	**PACIFIC COAST**	**LENGTH**
Massachusetts	192	California	840
New Hampshire	13	Oregon	296
New Jersey	130	Washington	157
New York**	127	* Alaska and Hawaii are not included in this activity.	
North Carolina	301		
Pennsylvania	0		
Rhode Island	40	** Includes the Atlantic Ocean coast of Long Island	
South Carolina	187		
Virginia	112		

THE EARTH'S GRID

5-1   Latitude As a Measurement Device (Instructor) 58

5-2   Constructing a Latitude Diagram 59

5-3   Finding Distances Between Places on Lines of Latitude 60

5-4   Determining Latitude and Distance on a World Map 61

5-5   Special Lines of Latitude That Define Earth's Zones (Instructor) 62

5-6   Understanding the World's Climate Zones 63

5-7   Longitude As a Measurement Device (Instructor) 64

5-8   Locating Places and Measuring Distance on Lines of Longitude 65

5-9   Using Latitude and Longitude to Find Locations and Distances 66

5-10 Longitude Helps Us to Tell Time 67

LATITUDE AS A MEASUREMENT DEVICE

Background

Maps of towns and cities are relatively easy to use if one wants to locate a place. But, even those maps are more useful if they have a grid of some kind that divides the space covered by the map into sections. Grids are a necessity in larger maps such as state, country, or world maps. How does one locate a place in an ocean, for example, if there is no grid for reference?

Place location is not the only use for a grid. Maps that show lines of latitude and longitude are, in essence, grids. Such maps are useful not only in place location, but also in matters of climate. In general, and not considering other variables, the farther one goes north or south of the Equator (0° latitude) the colder it becomes.

Following are some facts relative to latitude as a grid element.

1. Lines of latitude run east and west across a map or globe. They are circles and, as such, they have no specific beginning or end.

2. Lines of latitude are parallel to each other. Because of this fact, they are often referred to as "parallels."

3. The measurement of latitude starts at the Equator, which is 0° latitude. A place is either on the Equator, or south or north of the Equator.

4. The number of parallels shown on a map and their enumeration depends on the size of the area shown on the map. On a world map the parallels may proceed in units of ten. On a map of a small area, say New Jersey, the parallels may be spaced only 1° or 2° apart.

5. When identifying a particular line of latitude, designate it as either north or south of the Equator—for example, 8°N, or 30°S, and so on. In designating north or south, the abbreviation N for north or S for south is used. If a place is on the Equator no north or south designation is used.

In the diagram, places A and B are north of the Equator, place C is on the Equator, and place D is south of the Equator.

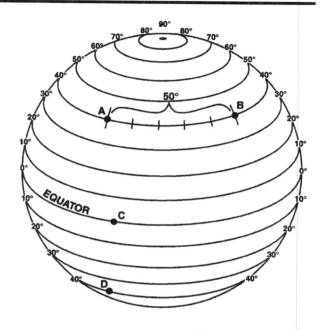

6. Lines of latitude vary in length. The Equator is 24,902 miles in length and is Earth's longest parallel. As the poles are approached, the lines of latitude shorten.

7. The lines of latitude are of different lengths, but they are all circles of 360°. Thus, when a particular latitude line is divided by 360°, we can determine the length of one degree on that particular line.

The diagram above tells that there are 50° between A and B. Note that the short vertical lines on the 40° line represent lines of longitude. The table below tells how many miles there are in one degree on the various lines of latitude. So, the distance between A and B on the 40° line of latitude is 2,650 miles (50° × 53 miles).

APPROXIMATE NUMBER OF MILES IN ONE DEGREE ALONG CERTAIN LINES OF LATITUDE			
Latitude	**Miles in 1 Degree**	**Latitude**	**Miles in 1 Degree**
Equator	70	50°N or S	45
10°N or S	69	60°N or S	35
20°N or S	65	70°N or S	24
30°N or S	60	80°N or S	12
40°N or S	53	* In common usage the Equator is said to be 25,000 miles in length.	

CONSTRUCTING A LATITUDE DIAGRAM

Lines of latitude, sometimes called *parallels*, are imaginary lines that circle the earth. The lines are numbered starting from the Equator. They are helpful in locating places. For example, a place can be said to be on the 40°N line, or the 20°S line, and so on. The lines are of different lengths as they approach the north and south poles.

To Do:

1. Number the lines on the diagram below. The lines are 10° apart. The Equator and the 10°S line and the 10°N line are already numbered to help you get started. Don't forget to put N for north and S for south at the end of the numbers. Be neat.

2. Now that you have numbered the lines, locate the following latitude lines on the diagram by writing its letter next to the appropriate dot. One place has already been done to help you get started.

A: 30°S	E: 50°S	I: 50°N
B: 20°N	F: 60°S	J: 40°S
C: 60°N	G: 70°N	K: 20°S
D: 10°S	H: 80°N	L: 30°N

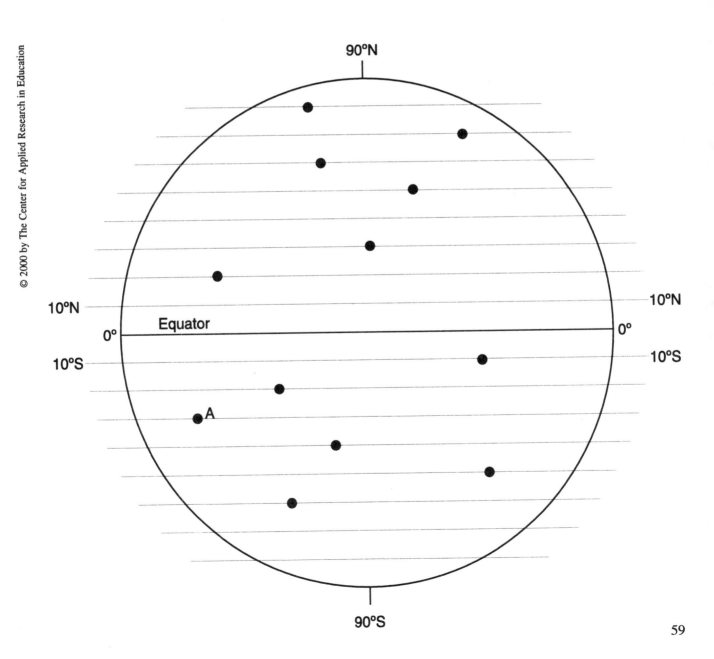

FINDING DISTANCES BETWEEN PLACES ON LINES OF LATITUDE

Some lines of latitude are much longer than others. The Equator, which is the 0° of latitude, is about 25,000 miles long. Much farther north or south, the 80° line is only about 4,300 miles long.

Every circle has 360°. So, if we divide the length of a line of latitude by 360°, we can determine how many miles there are in one degree of that particular line. The table below tells how many miles there are in every tenth line of latitude.

APPROXIMATE NUMBER OF MILES IN ONE DEGREE ALONG CERTAIN LINES OF LATITUDE			
Latitude	**Miles in 1 Degree**	**Latitude**	**Miles in 1 Degree**
Equator	70	50°N or S	45
10°N or S	69	60°N or S	35
20°N or S	65	70°N or S	24
30°N or S	60	80°N or S	12
40°N or S	53		

The diagram below shows several places that are located by dots and letters. To find the distances between the places, multiply the number of degrees between the dots by the number of miles in each degree on that particular line. For example, between dots A and B on latitude 80°N there are 55 degrees. According to the table, each degree on 80°N is equal to 12 miles. So, 55° × 12 miles equals a total of 660 miles between the two dots.

To Do:

Use the table and multiplication to find the miles between each of the following.

a.   A–B: _____ miles

b.   C–D: _____ miles

c.   E–F: _____ miles

d.   G–H: _____ miles

e.   I–J: _____ miles

f.   K–L: _____ miles

g.   M–N: _____ miles

© 2000 by The Center for Applied Research in Education

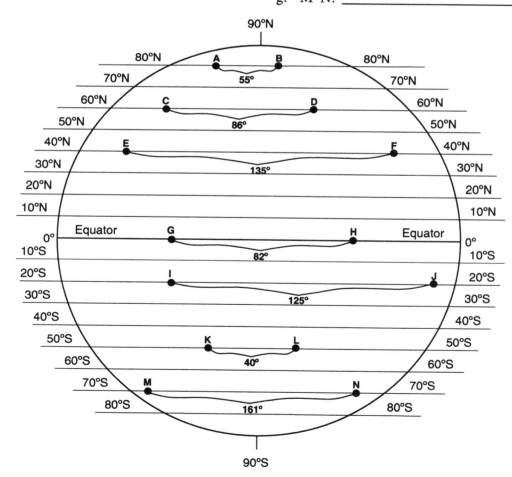

Name: _____ **Date:** _____

DETERMINING LATITUDE AND DISTANCE ON A WORLD MAP

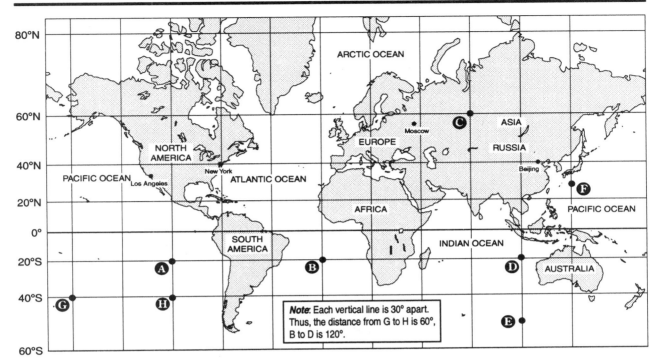

To Do:

1. What is the latitude of each of the following places? Be sure to include north (N) or south (S) in your answers.

If a place is not exactly on a line of latitude, make an estimate of its latitude. For example, if a place is midway between 40°N and 60°N, its latitude would be 50°N.

PLACE	LOCATION
New York	
Beijing (China)	
Los Angeles	
Moscow	
Ⓐ	
Ⓑ	
Ⓒ	
Ⓓ	
Ⓔ	
Ⓕ	

2. Follow the 20°S line across the map from west to east. What three continents does the line cross?

_____

_____

_____

3.a. One degree on the 40°S line of latitude is 53 miles. What is the distance from G to H?

_____ miles

b. One degree on the 20°S line of latitude is 65 miles. What is the distance from B to D?

_____ miles

4. Make your own latitude problem by printing J, K, and L at different places on the map. Then, with your teacher's permission, ask a classmate to tell you the latitude or approximate latitude at which each letter is printed.

SPECIAL LINES OF LATITUDE THAT DEFINE EARTH'S ZONES

Background

There are five imaginary, but very special, lines imposed on most world maps and globes. Their names and latitudinal positions follow: Equator, 0°; Antarctic Circle, 66½°S; Arctic Circle, 66½°N; Tropic of Capricorn, 23½°S; and Tropic of Cancer, 23½°N.

• The Equator divides the world into two halves, the Northern Hemisphere and the Southern Hemisphere.

• The Tropic of Cancer is the imaginary line over which the sun's rays are vertical on June 22. On this date the sun is directly overhead.

• The Tropic of Capricorn is the imaginary line over which the sun's rays are vertical on December 22. On this date the sun is directly overhead.

• The land and water south of the Tropic of Cancer and north of the Tropic of Capricorn are known as the Tropics or Tropical Zone. The weather is always warm or hot. The exception to this would be in mountains, which, because of their altitude, can be cold or even experience snow. In none of the lands north or south of the Tropics is the sun ever directly overhead; for them, the rays of the sun are always slanted.

• The Arctic Circle is the imaginary line that marks the beginning of the Arctic region. After June 22 the area from the Arctic Circle to the North Pole has continuous daylight; the sun does not set for six months. Then, beginning on December 22, the region is covered with darkness for six months. The area between the Arctic Circle and the North Pole is known as the North Frigid Zone.

• The Antarctic Circle is the imaginary line that marks the beginning of the Antarctic region. This region also has six months of darkness and six months of daylight. Here, the date when continuous daylight begins is December 22, and continuous darkness begins on June 22. These dates are directly opposite of those in the Arctic region. The area between the Antarctic Circle and the South Pole is known as the South Frigid Zone.

• The region between the Tropic of Cancer (23½°N) and the Arctic Circle (66½°N) is known as the North Temperate Zone. The region between the Tropic of Capricorn (23½°S) and the Antarctic Circle (66½°S) is known as the South Temperate Zone.

South of the Antarctic Circle

• No countries are included in the South Frigid Zone. The continent of Antarctica is completely within the Antarctic Circle except for the Antarctic Peninsula, which extends into the South Atlantic Ocean. Numerous islands are off the coasts of Antarctica.

• The continent nearest to Antarctica is South America. Cape Horn is about 600 miles north of the Antarctica Peninsula.

• The Vinson Massif (16,864') is the highest point in Antarctica. There are several other mountains there with elevations of more than 14,000'.

• In the Antarctic winter, the water surrounding the continent freezes over to make up an area of ice almost as large as the continent itself.

• In 1960 the coldest temperature ever recorded on Earth occurred in Antarctica: -127°F. This temperature was recorded in August, close to the middle of the Antarctic winter.

North of the Arctic Circle

• Three continents—Asia, Europe, and North America extend north of the Arctic Circle. Parts of Russia, Finland, Norway, Sweden, Canada, and the United States (Alaska) extend into the Arctic Zone.

• Airplanes frequently fly over the Arctic region on Great Circle routes from Europe to the United States.

• The Arctic region supports much more plant life than Antarctica does. During the summer season of continuous sun, a large variety of flowers such as poppies and bluebells flourish. Forests of stunted trees can also be found in the southern regions of the Arctic.

• Animals such as caribou and reindeer live in the southern Arctic. People who live in the region have herds of reindeer, which provide food, transportation, and furs.

• In winter, the temperature in the Arctic averages about -30°F. Some places in Siberia have had temperatures as low as -93°F. However, in summer, it can get very warm. Surprisingly, temperatures of 90°F have been recorded.

Suggestions for Teaching

The facing page may be shown via a transparency as the various facts on this page are explained.

Name: _____ Date: _____

UNDERSTANDING THE WORLD'S CLIMATE ZONES

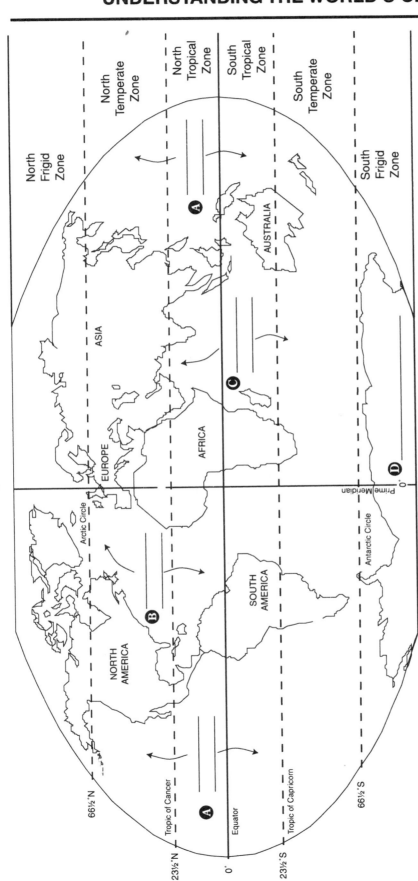

The map of the world shows the lines of latitude that are special. They mark the climate zones of Earth.

To Do:

1. Label the following on the map:
   Ⓐ Pacific Ocean
   Ⓑ Atlantic Ocean
   Ⓒ Indian Ocean

2. The southernmost continent is Antarctica. Label ANTARCTICA at Ⓓ on the map.

3. What line of latitude is the northern boundary of the South Frigid Zone? _____

4. What line of latitude is the southern boundary of the North Frigid Zone? _____

5. Between what two lines of latitude is the North Temperate zone? _____ and _____

6. Between what two lines of latitude is the South Temperate Zone? _____ and _____

7. What two lines of latitude are the northern and southern boundaries of the two Tropical Zones? _____ and _____

8. What line passes through the middle of the two Tropical Zones? _____

9. What two continents are completely south of the Equator? _____ and _____

10. What zone contains the most land? _____

LONGITUDE AS A MEASUREMENT DEVICE

Background

1. Lines of longitude differ from lines of latitude in several ways:

a. They are not parallel. The lines converge at the North Pole and the South Pole. They are farthest apart from each other at the Equator.

b. They are not significant indicators of climate.

c. They measure distances east and west of the Prime Meridian, which is 0° longitude. The Prime Meridian serves the same purpose as the Equator in terms of dividing the world into two halves, except that it divides the world into an Eastern Hemisphere and a Western Hemisphere. 0° to 180°W is the Western Hemisphere, and 0° to 180°E is the Eastern Hemisphere. 180°E and 180°W are the same line.

d. They are all the same length around Earth's circumference. Each line is approximately 25,000 miles in length. However, the same line has two different designations. The Prime Meridian, or 0° line of longitude, extends from the North Pole to the South Pole as the 0° line. When the line continues from the South Pole to the North Pole, it is known as the 180° line. Another example: the 30° line extends from the North Pole to the South Pole; on its return trip from the South Pole to the North Pole the same line is identified as the 150° line.

The diagram below illustrates how one longitude line is divided into two sections.

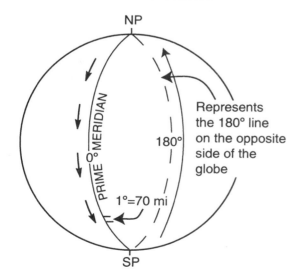

2. Lines of longitude in conjunction with lines of latitude are useful for locating places in the world. The place where an east-to-west latitude line crosses a north-to-south longitude line can be accurately identified. In identifying a location, its latitude is stated first and is followed by its longitude, as, for example, 40°N-75°W.

3. Every 24 hours, Earth rotates 360° on its axis. 360° divided by 24 hours equals an east-to-west movement of 15° each hour. Marking off globes and maps in 15° divisions makes it easy to tell sun-time. If it is 12:00 noon on the Prime Meridian (0°), it is 11:00 AM on the 15°W line, and it is 1:00 PM on the 15°E line. On the 30°W line it is 10:00 AM, and on the 30°E line it is 2:00 PM , and so on.

4. Earth is slightly flat at each of the poles. This makes the circumference of a line of longitude slightly less than the circumference of the Equator. In general use, however, the lines of longitude are considered to be 25,000 miles divided by 360°, which equals 70 miles per degree. This is unlike lines of latitude, which have fewer miles per degree as the circumference of the lines shorten as they approach the poles. *Note*: The diagram in the first column marks off 1° on a line of longitude as being 70 miles.

5. A Mercator map is often used in determining the location of places. It is accurate because the lines of latitude and longitude are vertically and horizontally straight. However, because lines of longitude are not shown as converging at the poles, the land and sea areas appear to be much larger than they are. For example, on Mercator maps, Greenland appears to be larger than South America. It would be helpful to student understanding to show a globe, which is the truest representation of Earth, in contrast to a printed Mercator map. Polar regions can then easily be seen to be distorted on the Mercator projection.

Name: _____ Date: _____

LOCATING PLACES AND MEASURING DISTANCE ON LINES OF LONGITUDE

Locating Places on Lines of Longitude

Lines of longitude are imaginary north-south lines that stretch from the North Pole to the South Pole. Figure 1 shows how longitude lines are usually numbered. The 0° line, also called the Prime Meridian, is where longitude lines begin. A place is either on the 0° line or east or west of it. Place A is on the 75°W line of longitude. Place B is on the 60°E line of longitude.

To Do:

On what line of longitude is each of the following places?

Place	Longitude	Place	Longitude
C		G	
D		H	
E		I	
F		J	

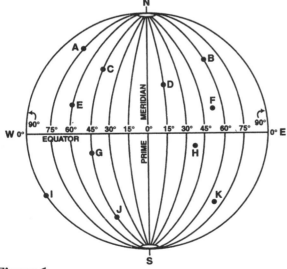

Figure 1

Note: You should realize that the diagram shows only one-half of the longitude lines. A flat map of the world would show that the lines extend 180°W and 180°E.

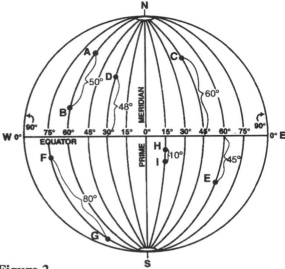

Figure 2

Mastering Distances on Lines of Longitude

Distances on lines of longitude are easy to measure. This is because one degree on every line of longitude is equal to 70 miles. If a place is 20°N of the Equator, it is 1400 miles north of the Equator. The multiplication problem is 20° × 70 miles per degree = 1400 miles.

Place A on figure 2 is located 50° north of place B. How far in miles are the two places from each other? Multiply 50° × 70 miles per degree. The answer is 3500 miles.

To Do:

1. How far north or south of the Equator is each of the following places?

Place C: _____ miles

Place D: _____ miles

Place E: _____ miles

2. How far apart in miles are the places listed below? The number of degrees that separate the two places is noted in the middle of the bracket.

a. F and G? _____ miles

b. H and I? _____ miles

Name: _____ **Date:** _____

USING LATITUDE AND LONGITUDE TO FIND LOCATIONS AND DISTANCES

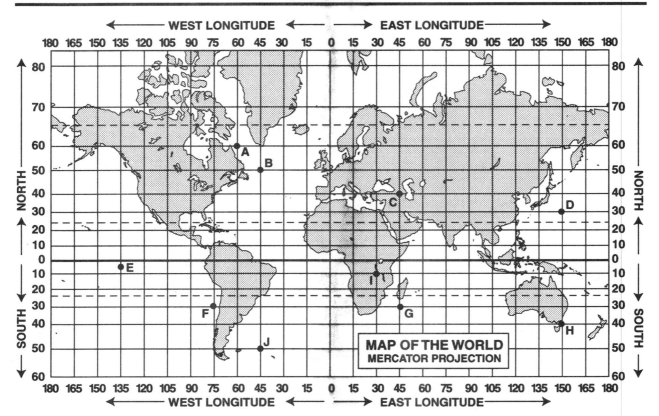

Latitude and longitude work together in locating places around the world. Where the horizontal lines of latitude cross the vertical lines of longitude are very definite locations. On the map, place A is on the 60°N line of latitude and the 60°W line of longitude. If you were writing this location you would write it like this: A is 60°N-60°W. Here is another example: B is 50°N-45°W.

To Do:

1. What are the locations of the following places?

Place	Location	Place	Location
C		G	
D		H	
E		I	

2. Use what you have learned about measuring distances with latitude and longitude in the following problems. Remember that every degree on a line of longitude is equal to 70 miles. If you count the degrees between B and J, you will find that they are 100 degrees apart. How many miles is that? Multiply 100° × 70 miles per degree, and you find that they are 7000 miles apart.

What is the distance in miles between

a. C and G? _____miles   b. D and H? _____miles

3. How many miles south of the Equator is

H? _____miles   I? _____miles   J? _____miles

4. Each degree on the 30°N or the 30°S line is equal to 60 miles. How many miles apart are F and G?

_____ miles

LONGITUDE HELPS US TO TELL TIME

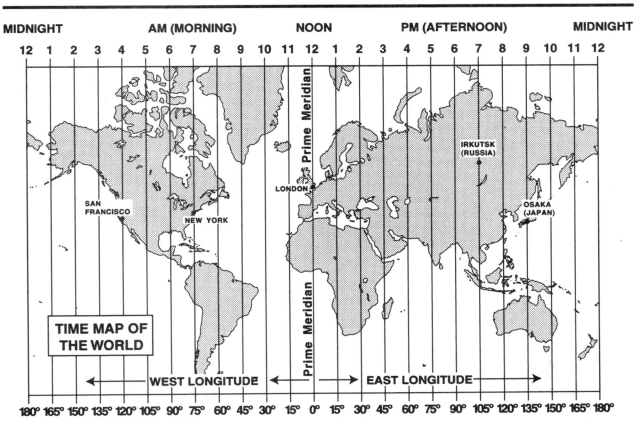

MIDNIGHT   **AM (MORNING)**   **NOON**   **PM (AFTERNOON)**   **MIDNIGHT**

12 1 2 3 4 5 6 7 8 9 10 11 12 1 2 3 4 5 6 7 8 9 10 11 12

TIME MAP OF THE WORLD

◄—— **WEST LONGITUDE** ——◄ **Prime Meridian** ►—— **EAST LONGITUDE** ——►

180° 165° 150° 135° 120° 105° 90° 75° 60° 45° 30° 15° 0° 15° 30° 45° 60° 75° 90° 105° 120° 135° 150° 165° 180°

© 2000 by The Center for Applied Research in Education

Earth turns on its axis 360° every 24 hours. If the 360° is divided by 24 hours, we find that Earth turns east toward the sun 15 degrees every hour. These facts help us to measure time.

Many years ago it was realized that there had to be a starting place for measuring time. It was decided that a line running through London, England, to the North Pole and the South Pole would be suitable. The line is called the ***Prime Meridian*** and is numbered 0°. Then, as the map shows, north-south lines were measured off starting with the Prime Meridian. The lines are measured every 15° west to 180° and every 15° east to 180°.

As the map tells you, when it is 12:00 noon on the Prime Meridian, it is 11:00 AM on the 15°W line. And, it is 1:00 PM on the 15°E line.

The above explanation helps you to know that if you live in New York and are eating lunch at 12:00 noon, students in San Francisco will probably be starting school at 9:00 in the morning. Then, in three hours, as Earth turns toward the sun, students in New York will be getting ready to go home from school at 3:00 in the afternoon when students in San Francisco are sitting down to lunch.

To Do:

1. When it is 12:00 noon on the Prime Meridian, what time is it at each of the following places? To find the answers, study the "times" across the top of the map.

Place	Time	Place	Time
London		Irkutsk (Russia)	
New York		Osaka (Japan)	
San Francisco			

2. How many hours difference in time is there between each of the places listed below?

a. San Francisco - New York: _____ hours

b. New York - London: _____ hours

c. London - San Francisco: _____ hours

d. London - Irkutsk: _____ hours

e. London - Osaka: _____ hours

f. Irkutsk - Osaka: _____ hours

UNDERSTANDING ALTITUDE

6-1   Altitude: Its Measurement and Significance (Instructor) 70

6-2   Understanding Altitude 71

6-3   The Use of Color to Show Elevation (Instructor) 72

6-4   Completing and Understanding a Color-Elevation Map 73

6-5   Understanding Contour Maps (Instructor) 74

6-6   Climbing Bell Mountain 75

6-7   Reading a United States Topographical Map 76

ALTITUDE: ITS MEASUREMENT AND SIGNIFICANCE

Background: Sea Level

The surfaces of the world's oceans are the starting points for measuring altitude. *Sea level* is another term for ocean surfaces. The expression *above sea level* is used in telling how high a place is above the level of the sea. Mt. Everest is 29,028' above sea level. Mt. Whitney, in California, is 14,494' above sea level. The graph illustrates how the two facts may be shown pictorially.

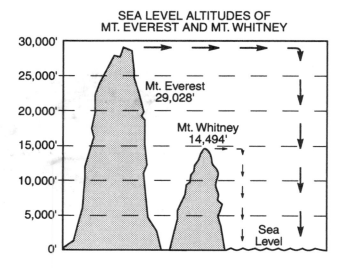

SEA LEVEL ALTITUDES OF
MT. EVEREST AND MT. WHITNEY

The term *below sea level* is used to tell how far places are below the surface of the sea. Death Valley, California, at its lowest point is 282' below sea level. The Mariana Trench, in the Pacific Ocean and east of the Philippine Islands, has a depth of 36,198'—that's more than 7,000' deeper than Mt. Everest is high.

The level of the sea can change, especially near coasts. As tides come in, the level rises; as tides go out, the level lowers. Sea level is the average between low and high tide. In any case, the difference between the two levels is not great, except, of course, in narrow inlets, where there may be as much as 30' difference between low and high tide.

Altitude and Temperature

Altitude has significant effects on temperature. The higher one goes, the lower the temperature. As a general rule, temperature drops three degrees for every 1000' increase in altitude. One may start to climb a 5000' mountain at a comfortable temperature of 65°F. However, at the peak of the climb the temperature may be 50°F. The wise mountain climber will be sure to have a warm jacket for use as the temperature grows colder.

Altitude and Vegetation

Anyone who has been in a vehicle ascending a mountain probably has noticed that as the road climbs the quantity and variety of vegetation changes. Trees become stunted as altitude increases, and they may not grow above the tree line, about 10,000' above sea level. Hardy crops such as oats, barley, and rye can grow at high altitudes as, for example, on the high plains of Bolivia.

Altitude and Oxygen

An increase in altitude brings about a decrease in oxygen. Hikers climbing high mountains find that breathing becomes more difficult as they ascend and that, as a result, they are easily tired for lack of oxygen. Just as oxygen must be provided for passengers in high-flying airplanes, oxygen is often required for high-climbing hikers.

Altitude and Airplanes

Air has weight. A column of air over a one-inch square patch at sea level and extending high into the atmosphere weighs about 15 pounds. The weight of air at 10,000' is much less—only 11 pounds per square inch; at 20,000' air has a weight of about 7 pounds per square inch.

It is the weight of air acting upon an instrument called an altimeter that translates the weight of air into altitude for pilots. Pilots must keep their airplanes well above the altitude of the land over which they are flying. Without altimeters, pilots would not know their altitude. In a fog or in the dark of night their airplanes could crash into mountains.

Name: _____ **Date:** _____

UNDERSTANDING ALTITUDE

A CROSS-SECTION OF A MAKE-BELIEVE LAND

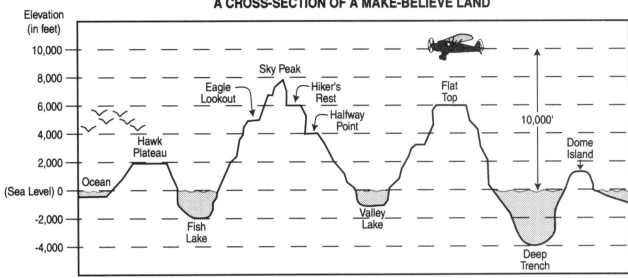

Another word for altitude is *elevation*. Both words tell us how high a place is above the surface of the sea, or sea level. If we read that a mountain has an elevation of 9000', we know that its peak is 9000' above sea level.

Note: The graph/diagram shows a *cross-section* of land and water. A cross-section is a view from the side. For example, the layers of a piece of cake seen from the side is also a cross-section.

To Do:

1. How many feet above sea level is

a. Hawk Plateau? _____

b. Eagle Lookout? _____

c. Sky Peak? _____

d. Hiker's Rest? _____

e. Flat Top? _____

f. Dome Island? _____

2. How many feet below sea level is the bottom of:

a. Fish Lake? _____

b. Valley Lake? _____

c. Deep Trench? _____

3. How many feet deeper is Deep Trench than Fish Lake? _____

4. How many feet above Flat Top is the airplane?

_____

5. If you were on Hiker's Rest, how many more feet would you have to climb to reach the top of Sky Peak? _____

6. Between what two levels of elevation are the birds flying? _____ and _____

7. How many feet higher in elevation is Sky Peak than Dome Island? _____

8. Imagine the airplane flying at its current elevation over Hawk Plateau. How many feet would the airplane have to come down to land there? _____

THE USE OF COLOR TO SHOW ELEVATION

Background

Reading elevation from a cross-section is easier than reading elevation from a map that uses a color scheme. The cross-section profile is closer to reality; the layers of elevation are readily interpreted. On a color elevation map, the map reader has to realize that different colors represent different degrees of elevation.

Reading a color map showing elevation may be explained and understood as follows.

1. Obtain a flat piece of cardboard and paint it blue. This represents the sea.

2. Obtain four round jar lids of different circumferences. Paint the sides and top of the largest lid green to represent 0 to 1000' above sea level. Place it on top of the blue cardboard.

3. Paint the sides and top of the next largest lid yellow to represent elevation from 1000' to 2000'. Place it on top of the green lid, and have students look at the side and top.

4. Paint the next largest lid brown to represent elevation from 2000' to 3000'. Place it on top of the yellow lid.

5. Paint the smallest lid red to represent elevation from 3000' to 4000'. Place it on top of the brown lid. (Figure 1)

6. Tell students they can see the elevation from the side and the top. Note that as they look down on the lids they get the impression of flatness.

7. Draw circles within circles on a piece of blue cardboard, and help students to understand that the jar lid tops and the drawn circles represent the same thing—elevation at varying levels. (Figure 2)

8. Place small objects (erasers, thumbtacks, etc.) on the jar lids. Have students tell the elevations on which the objects are resting. Then, do the same with the objects on the colored circles on the cardboard.

9. On the chalkboard draw a realistic profile of a mountain. Color ascending layers of green, yellow, brown, and red on the mountain. Help students to see that the board drawing shows the same things as the lids and the cardboard circles.

10. Finally, mold some clay into a mountain-like structure. Paint ascending circumferences of the mountain green, yellow, brown, and red. Place it on the blue cardboard. Help students to see that the lids, the drawn circles, the board drawing, and the model all show the same thing in different ways.

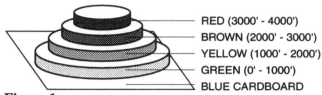

RED (3000' - 4000')
BROWN (2000' - 3000')
YELLOW (1000' - 2000')
GREEN (0' - 1000')
BLUE CARDBOARD

Figure 1

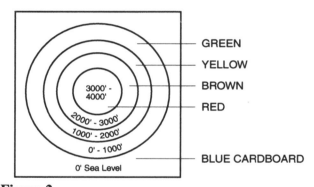

GREEN
YELLOW
BROWN
RED

3000' - 4000'
2000' - 3000'
1000' - 2000'
0' - 1000'
0' Sea Level

BLUE CARDBOARD

Figure 2

© 2000 by The Center for Applied Research in Education

Following are elements of color representation of elevation that will increase student understanding.

1. It is important to check the key to a map to determine what colors are used to show elevation. Different map makers may use different approaches to coloration. Also, colors may show different elevation extents. The figure above shows 1000' intervals. Some maps may show intervals of 2000' or more.

2. A map may show an area as being completely one color. However, within the elevation shown there may be deep valleys or high ridges that are just too small to show on a map section that encompasses several thousand square miles.

Name: _____   **Date:** _____

COMPLETING AND UNDERSTANDING A COLOR-ELEVATION MAP

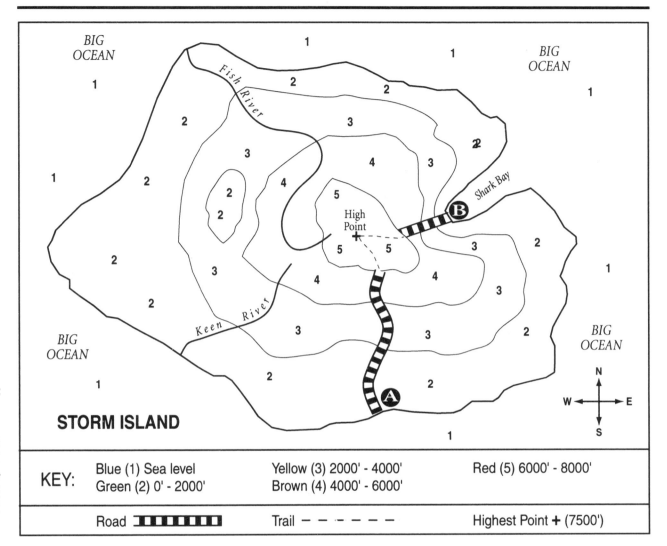

BIG OCEAN

BIG OCEAN

Fish River

Shark Bay

High Point +

Keen River

BIG OCEAN

BIG OCEAN

STORM ISLAND

N W E S

KEY:
Blue (1) Sea level
Green (2) 0' - 2000'
Yellow (3) 2000' - 4000'
Brown (4) 4000' - 6000'
Red (5) 6000' - 8000'

Road ▮▮▮▮▮▮▮   Trail – – · – – –   Highest Point + (7500')

On maps elevation may be shown by color or by numbers. A certain color or number represents a particular range of elevation. The important thing to do is to study the key of the map. Maps differ in the color or number scheme they use.

To Do:

1. What is the range of elevation at the source of the Keen River? _____

2. In what direction is the Keen River flowing?

___Southwest   ___Southeast

___Northwest   ___Northeast

3. Which river flows the greater distance in the 4000' - 6000' range of elevation?

___Keen River   ___Fish River

4. According to the key of the map and the map itself, what is the elevation of the highest point on Storm Island? _____ feet

5. Notice the two roads going from sea level to High Point.

a. Which road is steeper? _____ **Ⓐ**   _____ **Ⓑ**

b. Why couldn't a car go all the way up to High Point? _____

_____

6. Notice the small number 2 area within the number 3 area. Is this area a depression, or is it a high elevation? ___Depression   ___Elevation

7. Study the key to the map. It will tell you what colors to use to show elevation. Lightly color all the divisions on the map.

73

UNDERSTANDING CONTOUR MAPS

Background: Topographical Maps

The word *topography* has special meanings as related to maps. *Topo* as used in topography has to do with place, and *graphy* has to do with writing and depicting. Both parts of the word are derived from Greek. So, a topographical map is one that shows human-made and natural features of a particular area—usually in great detail. Such maps also contain elevations and the relative positions of places.

Topographical maps are indispensable tools for engineers, road builders, surveyors, and various state and national government agencies. In national defense, for example, topographical maps would be studied before an attack was launched on an enemy stronghold. Airplane pilots, especially those who fly single-engine planes, study topographical maps to plot their flights. Is there a radio tower that could be hidden in the fog? This is an example of the things pilots of small planes are concerned about that would be shown on the map.

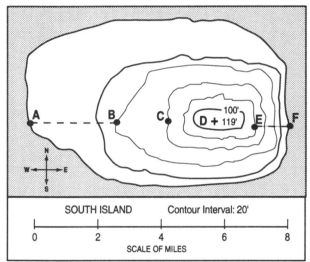

Figure 1

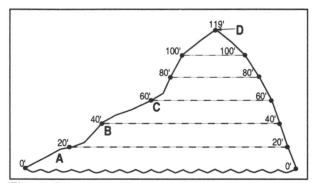

Figure 2

Suggestions for Teaching

1.   Make a transparency of Figures 1 and 2, and project it with the explanations below.

a.   The lines on topographical maps are called *contour lines*. They are imaginary lines, and every point on the line is the same height above sea level. If you could walk on a contour line you would never go up or down.

b.   Contour lines on maps are generally at intervals of 20'. The key to the map will state the contour interval.

c.   Sea level is considered to be the first contour line and is designated at 0'.

d.   The scale of miles is a tool that you can use to determine distance in a straight line, as an airplane would fly. The scale does not take into consideration that the walking distance from one point to another may be greater because of dips and rises in the road.

Explanation of Figure 1

1.   Point A is at sea level (0').

2.   Point B is at 40' above sea level.

3.   Point C is at 60' above sea level.

4.   Point D is somewhere between 100' and 119' above sea level. Topographical maps usually show the actual highest point and designate its location with a symbol.

5.   The distance from A to B is approximately three miles in a straight line. The land has reached an elevation of 40' in that three-mile distance.

6.   The distance from E to F is approximately one mile. The land has reached an elevation of 80' in that one-mile distance.

Explanation of Figure 2

The contour map, Figure 2, is shown in profile, or cross-section. Notice that the western slope, where the contour lines are relatively wide apart, has a gradual slope to the 60' contour line; then the grade becomes much steeper. The eastern side of South Island is much steeper, as indicated by the close contour lines. The rise from E to F is 80' in only one mile of distance.

CLIMBING BELL MOUNTAIN

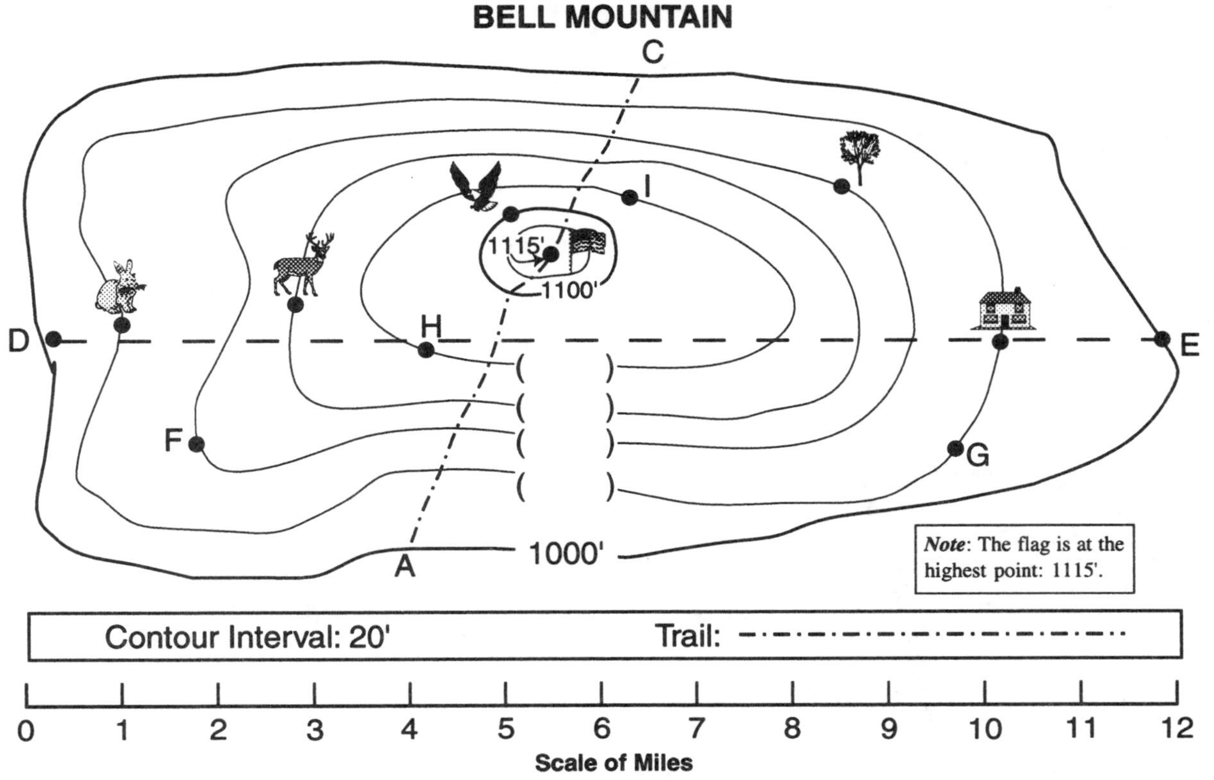

BELL MOUNTAIN

1115'
1100'
1000'

C
I
H
D
E
F
A
G

( )
( )
( )
( )

Note: The flag is at the highest point: 1115'.

Contour Interval: 20'	Trail: –·–·–·–·–·–·–·–

0   1   2   3   4   5   6   7   8   9   10   11   12

Scale of Miles

To Do:

1. Every point on a contour line is the same height above sea level. On the map above, the elevation distance from one contour to another is 20'.

   Write the number of each contour in the parentheses on each line.

2. Notice the symbols placed on the lines. The dot next to the symbol is the actual location of the symbol. What is the elevation of each of the symbols?

Symbol or Object	Elevation
a. House	
b. Tree	
c. Eagle	
d. Deer	
e. Rabbit	
f. Flag	

3. Which trail is shorter?

___A to Flag   ___C to Flag

4. Which trail is steeper?

___A to Flag   ___C to Flag

5. What is the distance in a straight line from:

a. D to E? _____ miles

b. H to the house?_____ miles

6. If a person were at H, could that person see another person at I? Why or why not?

_____

_____

_____

7. If you were to climb from the rabbit to the flag, how much higher would you be than when you started? _____ feet

8. Place a stick figure (⚲) on the trail from A to the flag on the 1020' contour line. Make a dot on the line at the proper place.

75

READING A UNITED STATES TOPOGRAPHICAL MAP

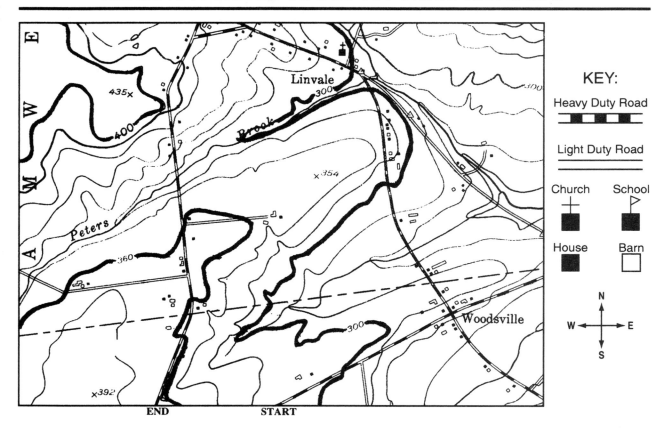

END   START

The map is a United States topographical map that shows a small part of New Jersey. The contour interval is 20'. Be sure to read the key before answering some of the questions.

To Do:

1. Begin a walk to Woodsville on the road that is above the word START.

a. What is the altitude of the first contour line you cross? _____ feet

b. Woodsville is located on a crossroad. Are there buildings on all four corners of the road?

___Yes   ___No

2. At Woodsville, turn to the left. Soon you cross the 300' contour line. How high is the high-point on your left (west) shown by an X?_____ feet

3. Continue your walk and reach Linvale. What kind of special building do you see at the corner?

___School   ___Church

4. You leave Linvale and make two lefts. As you walk south you are close to the 400' contour line. What is the elevation of the high point to the west?

_____ feet

5.a. Continue your walk and cross a brook. What is the brook's name? _____

b. What kind of road is the first you find on your left (east) after crossing the brook?

___Light duty   ___Heavy duty

c. What two kinds of building are at the end of that road? _____

d. What contour line crosses where the two roads meet?_____

6. Continue your walk until you reach another road on your right (west). Number the contour line just south of the 360' line as 380'. What is the elevation of the high point within the line you have just numbered? _____

GLOBAL MAPS AND POLAR MAPS

7-1   Map Projections: Some Pros and Cons (Instructor) **78**

7-2   Globe Maps and Lines of Latitude **79**

7-3   Working with Longitude I **80**

7-4   Working with Longitude II **81**

7-5   Latitude and Longitude on a Globe Map **82**

7-6   Polar Maps (Instructor) **83**

7-7   Working with Polar Maps I **84**

7-8   Working with Polar Maps II **85**

MAP PROJECTIONS: SOME PROS AND CONS

Background

Global maps offer certain advantages over flat maps.

1. Flat maps are not realistic in terms of shapes and sizes of places. A Mercator world map, for example, shows lines of longitude as perpendicular and parallel to each other; such lines would not converge at the poles. The result is that polar areas are greatly enlarged and misshaped. Greenland, for example, appears to be as large as South America. Antarctica appears to spread far east and west across the southern part of the map. The only parts of a Mercator world map that are reasonably accurate are the lands and waters in close proximity to the Equator.

2. It is difficult for inexperienced map readers to see the continuity of lands and waters on a flat world map. For example, the typical Mercator map shows the Pacific Ocean as "ending" in both the west and east. Even if one makes a cylinder of a flat map and, thus, has the ends of the Pacific Ocean meet, the polar sections are still left hollow.

3. It is difficult to understand polar routes of airplanes on a Mercator projection or on a modified world map on which the lines of longitude may be slightly curved so as to distort the polar regions less. It takes a great stretch of the imagination for young students to understand that airplanes fly over the North Pole when flying from Moscow to Seattle, Washington. A flat map "tells" students to fly west, or east, to reach Seattle from Moscow. Also, an

unknowing person finds it hard to believe that an airplane leaving New York for London would fly past Nova Scotia, Canada, and skirt Iceland rather than fly directly east. However, the Great Circle route as seen on a globe or global map tells us that the distance between points is much less than flying east or west between the same points as a Mercator map indicates.

Of course, there also are problems with global maps. At best, only one-half of Earth may be seen at one glance on a globe and/or global maps. A globe that spins is helpful; nevertheless, one-half of the world is always out-of-view.

Another problem that globes and global maps present concerns the reading of latitude and longitude. It is difficult to arrive at an accurate reading when a location is between two converging lines of longitude.

What are the answers to these problems? Students should learn advantages and disadvantages of flat maps such as the Mercator and the limitations of globes and global maps. One kind of map will answer certain questions and another kind of map will answer other questions. Therefore, it is important that instructors give appropriate attention to map study.

Suggestions for Teaching

Make transparencies of the two maps below. Point out the advantages and disadvantages of each projection. Stress, however, that a globe or global map portrays the most accurate shapes, sizes, and distances. All other maps are distorted in one way or another.

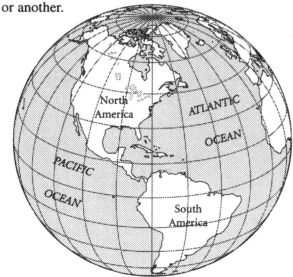

WEST LONGITUDE   EAST LONGITUDE

180° 165° 150° 135° 120° 105° 90° 75° 60° 45° 30° 15° 0° 15° 30° 45° 60° 75° 90° 105° 120° 135° 150° 165° 180°

GLOBE MAPS AND LINES OF LATITUDE

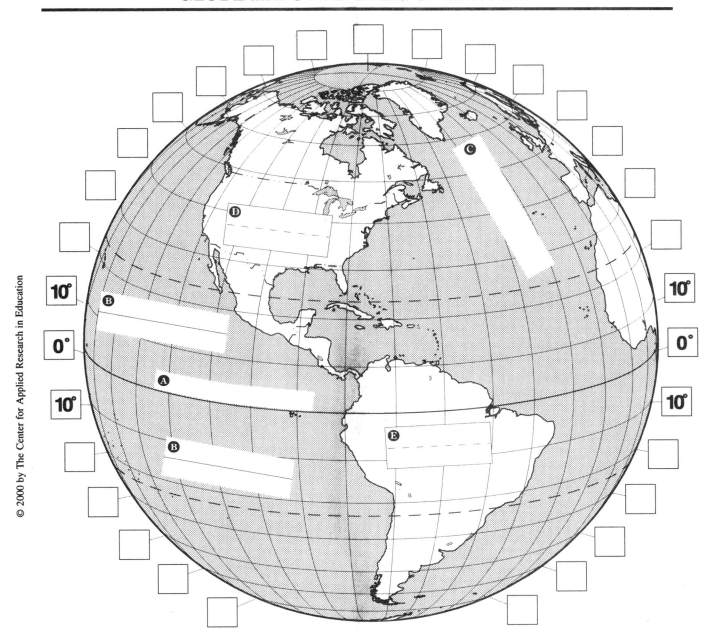

Because a globe map is a flat representation of a sphere, it can show only one-half of the world at a time. And the edges of the map are hard to read. It takes several different globe maps to show all the places clearly. Notice that the globe on this page is tilted so that you can see the North Pole.

To Do:

1. Make the globe more understandable by labeling the following lines of latitude. Start at the Equator, which is 0° latitude. In multiples of 10°, label the other lines of latitude. Print neatly in the boxes on both sides of the map. The 0° line and the 10° line have been done to help you get started.

2. Label the following:

Ⓐ Equator

Ⓑ Pacific Ocean (both places)

Ⓒ Atlantic Ocean

Ⓓ North America

Ⓔ South America

79

Name: _____ **Date:** _____

WORKING WITH LONGITUDE I

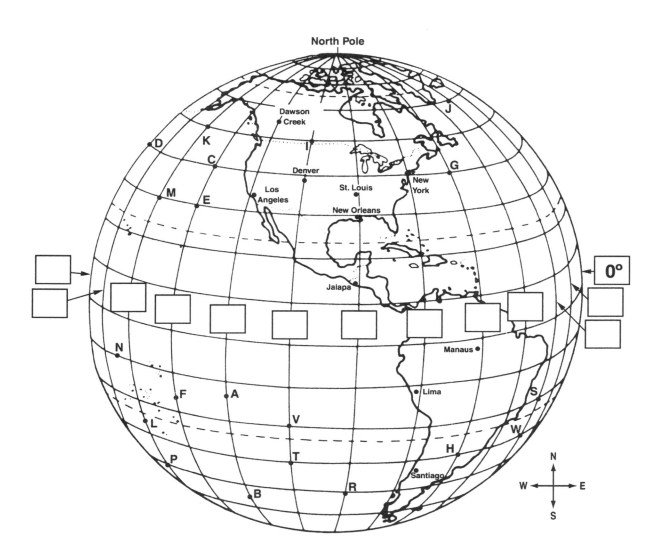

North Pole

Dawson
Creek

D   K   I   J

C   G

Denver   New
York

M   E   St. Louis

Los
Angeles   New Orleans

0°

Jalapa

Manaus

N   Lima   S

F   A   W

L   V

P   H

T

B   Santiago   R

N
W ← → E
S

GLOBAL MAP OF THE WESTERN HEMISPHERE

WORKING WITH LONGITUDE II

The lines of longitude on the globe map on the facing page are curved to give an impression of a sphere. However, if you were to examine a real globe you would see that the lines are actually straight north and south. Notice that all of the lines meet at the North Pole.

To Do:

1. The lines of longitude should be numbered. All of the lines are west of the Prime Meridian, or 0°. And all the lines are 15° apart on the Equator. Starting with the line of longitude numbered 0°, write the number of each line in the box. The numbers proceed west as follows: 15°, 30°, 45°, 60°, 75°, 90°, 105°, 120° 135°, 150°, 165°, 180°.

2. Letter A is directly north of letter B, even though it may not appear so. Every place on a particular line of longitude is directly north or south of every other place on the line. Similarly, every place on a particular line of latitude is directly east or west of every other place on the line: Place C, for example, is directly west of Place D, even though the line is curved.

a. What letter is directly south of each of the following letters?

K ___   I ___   G ___

E ___   M ___   S ___

b. What letter is directly west of each of the following letters?

G ___   H ___   I ___

R ___   E ___   C ___

3. On what line of longitude is each of the following places? Always print W for west after the numbers.

PLACE	LONGITUDE	PLACE	LONGITUDE
A		G	
B		H	
C		I	
D		J	

4. What line of longitude is nearest to each of the following cities?

CITY	LONGITUDE
Dawson Creek	
New York	
St. Louis	
Los Angeles	
New Orleans	
Denver	
Jalapa	
Manaus	
Santiago	
Lima	

LATITUDE AND LONGITUDE ON A GLOBE MAP

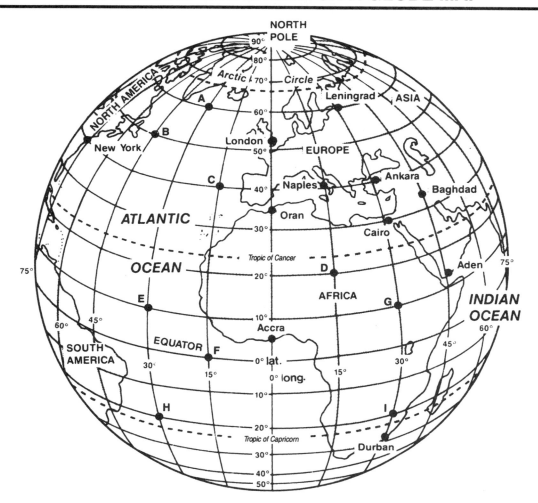

© 2000 by The Center for Applied Research in Education

Reading latitude and longitude on a globe map may be more difficult than reading locations on a flat map such as the Mercator map. The globe map has curved lines, and the Mercator map has straight lines. However, with practice you can become quite expert in using a globe for locations.

To Do:

The lines of latitude and longitude are numbered. The 0° line (Prime Meridian) is the dividing line between east and west. Notice that the line goes through London, England. Everything west of the Prime Meridian up to 180° is west longitude; everything east up to the 180° line is east longitude.

Everything north of the Equator is north latitude; everything south of the Equator is south latitude.

What is the latitude and longitude of the following places? Be sure to put W for west or E for east and N for north or S for south for all of the locations.

PLACE	LATITUDE/ LONGITUDE	PLACE	LATITUDE/ LONGITUDE
New York	40°N-75°W	A	
Naples		B	
Cairo		C	
Leningrad		D	
Ankara		E	
Baghdad		F	
Aden		G	
Durban		H	
Accra		I	

POLAR MAPS

Background

Polar maps may be completely new to some of your students. However, even those students who have experienced such maps will gain more knowledge and skill from the activity that follows.

Following are some aspects of polar maps that can be conveyed to students.

1. A polar map shows only one hemisphere of Earth—either the northern or the southern hemisphere. This is because either the North Pole or the South Pole is in the center of the map. This may come as a surprise to students; many think of north as being at the top of a map and south as being at the bottom.

2. On a north polar map all lines of longitude are north-to-south lines. The opposite is true on a south polar map where all lines of longitude are south-to-north lines. At first it is difficult for students to realize these facts. If this is the case, display Figure 1, which shows the 135° line and the 180° line as going "up". They might ask, "How can that be south?" If this question arises, call their attention to the longitude lines on a real globe—first from a side view, and then from a polar view. They should be able to understand that the lines on the diagram only appear to go toward the north.

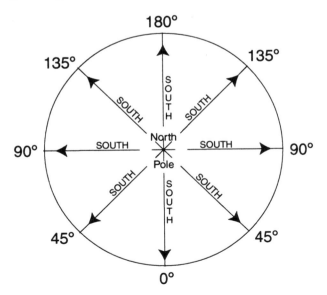

Figure 1

3. Lines of latitude on polar maps offer the same kind of problem as lines of longitude; that is, appearances are deceiving. A westward around-the-world air route gives the appearance of going west at the start. However, as the route approaches the 180° line of longitude, it appears to be going north. Then, as it passes the 180° line, it appears to be going south. As the route approaches the start of the flight, it appears to be going west once more. Again, show on a real globe that the route is always toward the west. Figure 2 shows the routes of two airplanes.

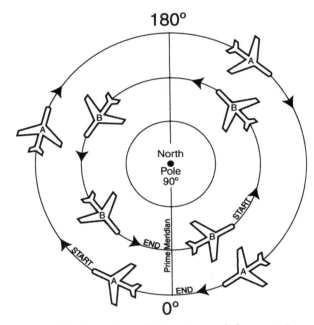

Airplane A is flying toward the *west*.
Airplane B is flying toward the *east*.

Figure 2

4. True airplane routes may be readily seen if they are drawn on a polar projection. An airplane route from New York to Moscow drawn on a Mercator map gives an erroneous impression of the way airplanes would actually fly. The Mercator map route takes an airplane over Ireland, France, Germany, and Poland.

The true route, as drawn on a polar map, shows that an airplane would fly over Greenland, Norway, Sweden, and Finland and, in the process, save hundreds of miles of flying. Such routes are known as Great Circle routes.

WORKING WITH POLAR MAPS I

The polar map on the facing page is not like most flat maps. You should think of yourself as being far out in space looking down at the North Pole. The land and waters of the Northern Hemisphere would be spread before you.

To Do:

1. The lines of longitude are the straight lines on the map. They should be numbered.

a. Start at the line numbered 0°, which is the Prime Meridian. Then, write the numbers of the other lines in the boxes. The numbers proceed to the east as the arrow indicates. Each line is 20° from the next line, as follows: 20°E, 40°E, 60°E, 80°E, 100°E, 120°E, 140°E, 160°E. The 180° line is already numbered.

b. Once again start at the line numbered 0°, but this time the lines are numbered to the west of the Prime Meridian. Number the lines as follows: 20°W, 40°W, 60°W, 80°W, 100°W, 120°W, 140°W, 160°W. The 180° line is already numbered.

2. Even though it looks as though the lines of longitude run in many directions, all the longitude lines run north-south and go all the way around the globe. Again, if you look at a real globe you will see that this is true.

What is the numbered longitude line that is nearest to:

a. New York? _____

b. Moscow? _____

c. Beijing? _____

3. The circle on the outside of the map represents the Equator. On the blank line between the 40°W line and the 60°W line write EQUATOR, 0°.

4. The circles are latitude lines that go east-west or west-east around the world. Again, if you look at a real globe you will see that this is true.

What is the numbered latitude line nearest to:

a. New York? _____

b. Moscow? _____

c. Beijing? _____

5. If you were to fly east or west along the 40°N line of latitude, you would eventually reach Beijing. However, there is another way that is *much* shorter.

a. With your ruler draw a heavy and dashed line connecting the dots that represent Beijing and New York. The route would take you very close to the North Pole. This route is called the Great Circle route.

b. The Great Circle route to Beijing is 6,684 miles.

c. The route that you would follow if your airplane flew west from New York to Beijing along the 40°N line is 10,050 miles.

How many miles are saved by flying the Great Circle route rather than flying the 40°N route?

_____ miles

6. What is the latitude and longitude of the following places shown by letters on the map?

A _____

B _____

C _____

D _____

7. In what direction are each of the airplanes flying? Draw a circle around the correct direction.

Airplane E:	North	East	South	West
Airplane F:	North	East	South	West
Airplane G:	North	East	South	West
Airplane H:	North	East	South	West

8.a. Locate the 100°W line of longitude. Follow it to the North Pole. What longitude number does the 100°W longitude line take from the North Pole onward? _____

b. What longitude number does the 120°E longitude line take from the North Pole onward? _____

Name: _____ Date: _____

WORKING WITH POLAR MAPS II

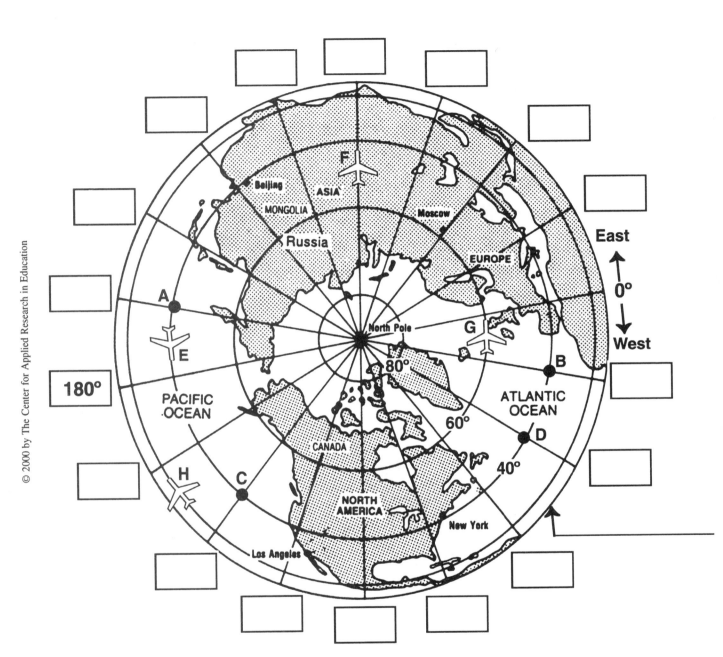

© 2000 by The Center for Applied Research in Education

POLAR PROJECTION OF THE NORTHERN HEMISPHERE

SECTION 8

GRAPHS

8-1 Reading Picture Graphs (Instructor) 88

8-2 A Picture Graph Shows a Farmer's Apple Harvest 89

8-3 A Picture Graph of a Family's Jam-Making Activities 90

8-4 A Picture Graph Showing Cotton Production in the United States 91

8-5 Reading Bar Graphs (Instructor) 92

8-6 The United States' Greatest Oil-Producing States 93

8-7 Comparing the Highest Points of Continents 94

8-8 Making and Reading Circle Graphs (Instructor) 95

8-9 Comparing Quantities and Sizes on Circle Graphs 96

8-10 Making and Reading Square Graphs and Single-Bar Graphs (Instructor) 97

8-11 A Square Graph Compares South America's Countries 98

8-12 Two Single-Bar Graphs: The Great Lakes and Fishing 99

8-13 Making and Reading Line Graphs (Instructor) 100

8-14 Understanding Tornadoes 101

8-15 Completing a Line Graph of Population Growth 102

READING PICTURE GRAPHS

Background

There are five main kinds of graphs: picture, bar, line, square, and circle. The most simple graphs are pictorial; that is, they show quantities by means of symbols. Each symbol is assigned a certain value. The reader then has to count the symbols on a line and multiply that number by the value assigned to the symbol. For example, a line that shows five symbols, say wheat, with each symbol having a value of six tons represents a total of 30 tons of wheat.

If the line showed four whole symbols of wheat and one half symbol, it would represent—using 6 tons as the value of each symbol—27 tons of wheat.

Younger learners who do not have skill in multiplying should read and/or make picture graphs that assign a value of one, possibly two, to each symbol. For example, two symbols on a line, each with an assigned value of 2, results in a total quantity of 4. Fourth graders who know the multiplication tables and who know how to solve multiplication problems should have little trouble understanding most picture graphs.

It is more difficult to make picture graphs than to read them. Therefore, the first picture graphs that follow are to be read, whereas the incomplete graphs that follow are to be completed by students.

It is important and helpful to use symbols in picture graphs that offer a resemblance to what is being graphed. In designing a picture graph that is concerned with automobiles, for example, a silhouette of an automobile would be an appropriate symbol.

In a graph concerned with houses, a simple house outline would be suitable and easily recognized.

Some Additional Considerations

1. Most often, picture graph symbols are arranged in horizontal rows as compared to vertical rows. In either case, the rows of aligned symbols resemble the bars on a bar graph. Thus, the reader is able to make a quick judgement of the relationships between and general trend of the graphed elements.

2. The title of the graph should clearly tell the main idea of the graph and, if appropriate, the years that the graph represents.

3. The quantity that each symbol represents should be stated at the bottom of the graph. It probably is easier for students to interpret symbols that represent even numbers (2, 4, 6, 8, etc.) than uneven numbers (3, 5, 7, 9, etc.). This is especially true if a part of a symbol is to be read or shown on a picture graph.

4. Whenever possible, the source of the information shown in the graph should be noted in relatively small print at the bottom of the graph.

Student Involvement

Question 8 of the activity page will require students to think beyond the factual information contained in the graph. No doubt, the answers the students suggest will differ. However, from the composite answers some significant speculation will emerge. (See suggested answers to question 8 in Answer Key.)

Name: _____   Date: _____

A PICTURE GRAPH SHOWS A FARMER'S APPLE HARVEST

FARMER JONES' APPLE HARVEST: 1991 - 2000

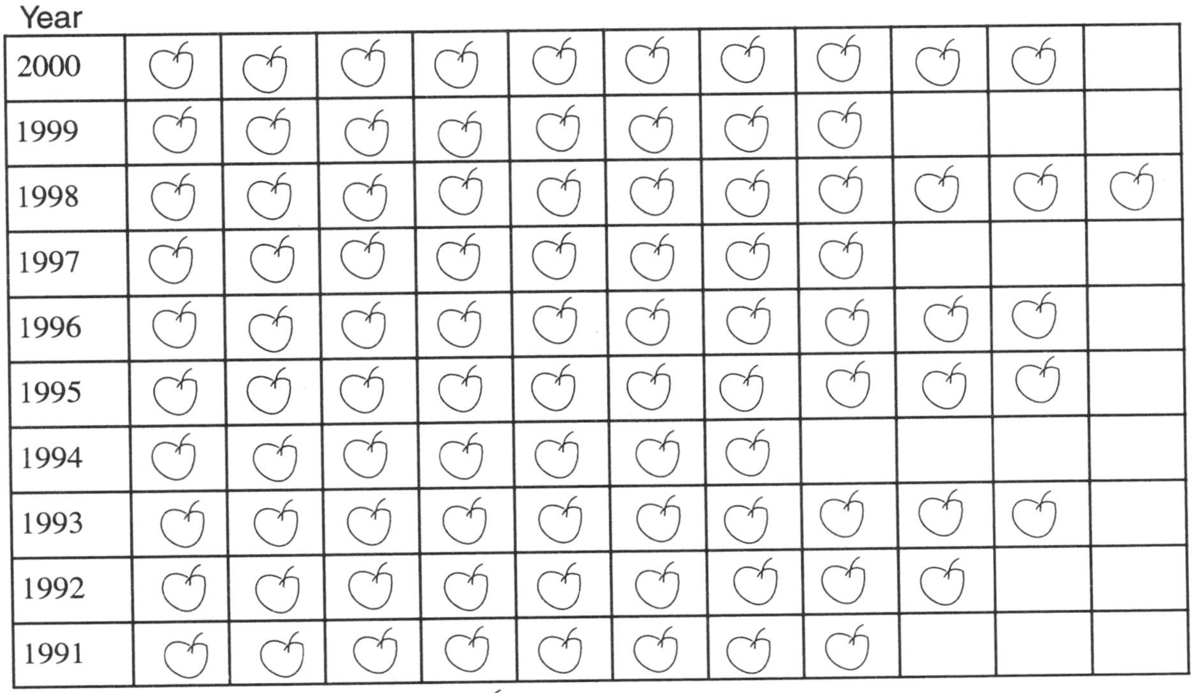

Year

Year											
2000	🍎	🍎	🍎	🍎	🍎	🍎	🍎	🍎	🍎	🍎	
1999	🍎	🍎	🍎	🍎	🍎	🍎	🍎	🍎			
1998	🍎	🍎	🍎	🍎	🍎	🍎	🍎	🍎	🍎	🍎	🍎
1997	🍎	🍎	🍎	🍎	🍎	🍎	🍎	🍎			
1996	🍎	🍎	🍎	🍎	🍎	🍎	🍎	🍎	🍎	🍎	
1995	🍎	🍎	🍎	🍎	🍎	🍎	🍎	🍎	🍎	🍎	
1994	🍎	🍎	🍎	🍎	🍎	🍎	🍎				
1993	🍎	🍎	🍎	🍎	🍎	🍎	🍎	🍎	🍎	🍎	
1992	🍎	🍎	🍎	🍎	🍎	🍎	🍎	🍎			
1991	🍎	🍎	🍎	🍎	🍎	🍎	🍎	🍎			

Each 🍎 = 20 bushels of apples

To Do:

1. How many *bushels* of apples did Farmer Jones pick each year?

YEAR	BUSHELS	YEAR	BUSHELS
2000		1995	
1999		1994	
1998		1993	
1997		1992	
1996		1991	

2. In what year was the harvest of apples the greatest? _____

3. In what year was the harvest of apples the lowest?

_____

4. How many fewer bushels were picked in the worst year than in the best year? _____

5. What is the total number of bushels harvested for all the years listed on the graph? _____

6. If each bushel weighed 40 pounds, how many pounds of apples did Farmer Jones harvest in the year 2000? _____

7. Imagine that Farmer Jones sold each bushel of apples for $45.00. How much money did he receive for the sale of his year 2000 crop?

_____

8. Try to think of a reason why Farmer Jones' 1994 crop was less than his 1993 and 1995 crop.

_____

_____

_____

_____

_____

Name: _____ Date: _____

A PICTURE GRAPH OF A FAMILY'S JAM-MAKING ACTIVITIES

The graph grid at the bottom of the page is an incomplete picture graph. As you can see, no titles, no symbols, and no quantities are listed. Your task in this activity is to take the information that is printed in the box and arrange it on the grid that has been provided.

Here are the steps you should follow:

1. Think of a title that tells in a few words what the graph is trying to show.

2. The items that are going to be shown on the graph should be easily recognized. For example, a graph about endangered eagles could use a simple outline of an eagle as its symbol.

3. Decide on a value for each symbol. For example, a picture graph that is concerned about oil production could use barrels for symbols. How many would one barrel represent? Whatever value is assigned to each symbol should be clearly shown on the graph, usually at the bottom.

4. Draw as many symbols as are necessary on the graph.

Mr. Smith and his family love peaches. The property the family owns has twenty peach trees that were planted many years ago. Each year, in September, the family picks peaches from the tree. The family then takes the peaches and makes jam from them, which they give to their friends. Each year the family keeps a record of how many two-pound jars of jam they make. Here is the record:

Jars of Jam Made

1994: 40	1997: 55
1995: 50	1998: 60
1996: 60	1999: 65

Suggestions for Completing the Graph Below

• An outline of a jar could represent five jars of jam. This would mean, for example, that eight symbols would be drawn on the 1994 line of the graph.

• Divide each year's production by five. Then draw the proper number of jars on each year's line.

Title: _____
_____

Year

1999														
1998														
1997														
1996														
1995														
1994														

JARS OF PEACH JAM

Each symbol (🫙) equals _____ jars of jam

© 2000 by The Center for Applied Research in Education

Name: _____ Date: _____

A PICTURE GRAPH SHOWING COTTON PRODUCTION IN THE UNITED STATES

Bales of Cotton Produced in Nine Leading States*

STATE

Texas	⊞	⊞	⊞	⊞	⊞	⊞	⊞	⊞	⊞	⊞
California	⊞	⊞	⊞	⊞						
Georgia	⊞	⊞	⊞	⊞						
Mississippi	⊞	⊞	⊞	⊞						
Arkansas	⊞	⊞	⊞							
Louisiana	⊞	⊞								
North Carolina	⊞	⊞								

- Each symbol (⊞) represents 500,000 (1/2 million) bales of cotton.
- Each bale of cotton weighs 480 pounds.
- The number of bales in each state has been rounded to the nearest 500,000.

World Almanac, 1999

To Do:

1. How many million bales of cotton were produced in each of the following states?

Texas: _____ California: _____

Georgia: _____ Mississippi: _____

Arkansas: _____ Louisiana: _____

North Carolina: _____

2. How many times greater was cotton production in Texas than Louisiana? ___3 ___4 ___5

3. Notice that the two bottom rows of the graph have been left blank. Complete the two rows by printing two more state names—Arizona and Tennessee—and drawing two symbols for Arizona, and one for Tennessee.

4. The map shows the location of the nine states that were leading cotton producers. The table tells you the names of the numbered states.

a. What four states border the Gulf of Mexico?

_____

_____

_____

_____

b. What state borders the Atlantic Ocean?

_____

c. What state borders the Pacific Ocean?

_____

d. What two states, along with Arizona, are completely inland?

_____

_____

91

READING BAR GRAPHS

Background

Bar graphs can give general impressions at a glance, and a closer examination can reveal more specific information.

Most bar graphs consist of two or more bars placed on a grid. The bars may extend from either the vertical axis or the horizontal axis. One axis indicates quantities, and the other indicates what is being graphed—years, state populations, etc. In essence, bar graphs show comparisons.

The quantity axis should be divided into equal units of measurement, usually even units of measurement such as 2, 4, 6, 8 or 10, 20, 30, etc. This makes it easier to read a quantity that may, for example, extend halfway into a division. The heading of the axis from which the bars extend should be specific: years, population, births, etc. A general rule applicable to most bar graphs is that there should be no more than ten enumerated units.

It is important that the quantity axis start at zero. This will make the bars proportional in length. The following illustrations show how bar graphs that do not start at zero render erroneous impressions.

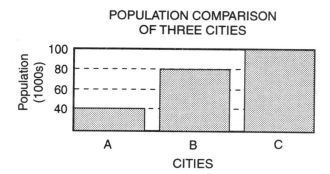

At first glance it seems that City B is three times greater in population than City A. However, this is a false impression, as the following correct graph shows. In actuality, City B has only twice the population of City A.

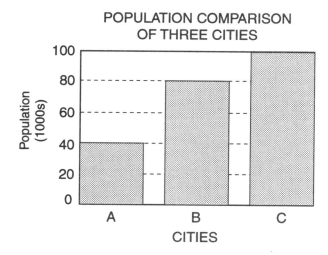

Please note that, if the quantity axis contains items with three or more zeros, some of the zeros can be eliminated. However, there should be an explanatory note on the graph. An example follows:

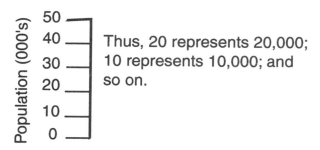

Thus, 20 represents 20,000; 10 represents 10,000; and so on.

Titles of graphs are important. They should be brief yet still convey the main topic of the graph. Time, place, and other factors should be very explicit.

Name: _____   Date: _____

THE UNITED STATES' GREATEST OIL-PRODUCING STATES

THE USA'S SEVEN GREATEST OIL-PRODUCING STATES
(In a Recent Year)

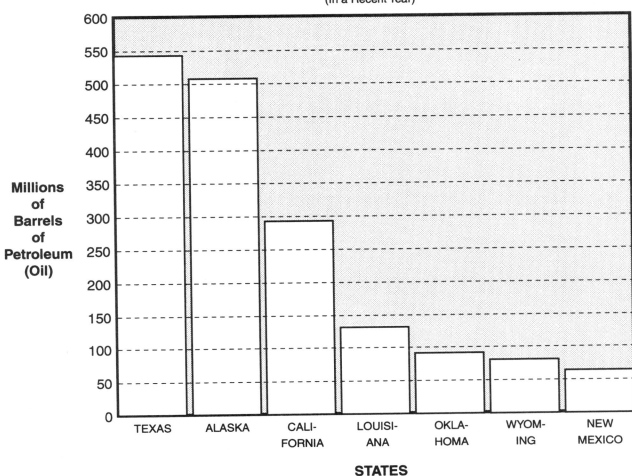

STATES

To Do:

1. To make the bar graph easier to read, do the following:

a. Print the numbers listed below on the lines at the tops of the bars. *Note*: You should realize that the numbers represent millions of barrels. Example: The 543 on top of the Texas bar represents 543 million (543,000,000) barrels of oil.

    Texas: 543
    Alaska: 510
    California: 282
    Louisiana: 132
    Oklahoma: 85
    Wyoming: 75
    New Mexico: 64

2. Draw diagonal lines (/////) in all the bars.

3. Does the combined oil production of Louisiana, Oklahoma, Wyoming, and New Mexico equal the oil production of Texas?

_____ Yes _____ No

4. How many more barrels of oil does California produce than Louisiana?

_____ million barrels

5. What is the total number of barrels of oil produced by the seven states?

_____ million barrels

6. Check on a map: Are most of the oil-producing states west of the Mississippi River?

_____ Yes _____ No

Name: _____ Date: _____

COMPARING THE HIGHEST POINTS OF CONTINENTS

THE HIGHEST POINT ON EACH CONTINENT

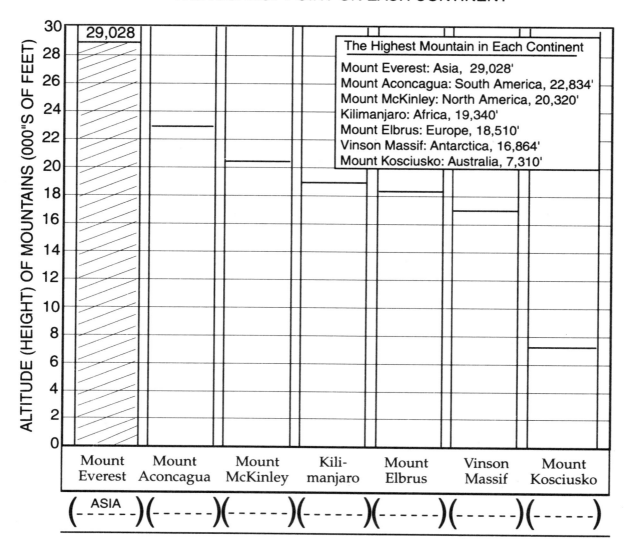

ALTITUDE (HEIGHT) OF MOUNTAINS (000'S OF FEET)

29,028

The Highest Mountain in Each Continent

Mount Everest: Asia, 29,028'
Mount Aconcagua: South America, 22,834'
Mount McKinley: North America, 20,320'
Kilimanjaro: Africa, 19,340'
Mount Elbrus: Europe, 18,510'
Vinson Massif: Antarctica, 16,864'
Mount Kosciusko: Australia, 7,310'

Mount Everest — Mount Aconcagua — Mount McKinley — Kili-manjaro — Mount Elbrus — Vinson Massif — Mount Kosciusko

(ASIA)(------)(------)(------)(------)(------)(------)

To Do:

1.a. The height of each mountain is shown by the heavy line at the top of each bar.

b. Fill in the bars with diagonal (////) lines.

c. At the top of each bar, print the height it represents. See the table for the heights.

d. At the bottom of each bar, print the name of the continent in which each mountain is located. Print in the parentheses (   ).

Note: Mt. Everest has been done to help you get started.

2. A mile is 5280' long. How many feet less than 5 miles is Mount Aconcagua?

_____ feet

3. How many feet less than Mount McKinley is Mount Kosciusko?

_____ feet

4. How many feet higher than Mount Aconcaqua is the combined height of Vinson Massif and Mount Kosciusko?

_____ feet

MAKING AND READING CIRCLE GRAPHS

Background

There are three kinds of *area* graphs: circle, square, and single bar. In these graphs, each element of whatever is being graphed is given a portion of the whole. In all three of these graphs, percent is a factor.

Following are some facts and procedures relative to making circle graphs.

1. Every circle has 360°.

2. What percent of the whole (360°) does each element to be graphed possess? Determining the percent becomes a mathematics problem. An example of how percent is determined follows:

a. A shopper spent $180, as follows: jacket, $90; shoes, $45; gloves, $36; handkerchiefs, $9.

b. What percent of the $180 is each of the purchases?
$$\$90 \div \$180 = 50\%$$
$$\$45 \div \$180 = 25\%$$
$$\$36 \div \$180 = 20\%$$
$$\$9 \ \div \$180 = \ 5\%$$

3. How are the percents converted to degrees of a circle?
$$50\% \text{ of } 360° = 180°$$
$$25\% \text{ of } 360° = 90°$$
$$20\% \text{ of } 360° = 72°$$
$$5\% \text{ of } 360° = 18°$$

4. How are the four degree segments plotted on a circle?

a. Draw a radius from the center of the circle to the top of the circle.

b. Lay the straight edge of a protractor along the radius.

c. Make a mark at 180° on the circle's circumference.

d. Draw a straight line from the mark to the center. The segment produced (jacket) is 50% of the circle.

e. For the next segment, lay the protractor edge against the second radius. Make another mark at 90° on the circumference of the circle. The second segment produced (shoes) is 25% of the circle.

f. Follow step *e* for the remaining two segments, except make marks at 72° and 18°, respectively.

g. Identify the segments as seen in the diagram in the next column.

h. Provide a title for the graph.

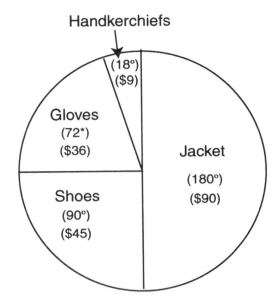

How $180 Was Spent
on Clothes

Suggestions for Teaching

1. Bring student attention to the segments of circle graphs mentioned in items 1-3 on this page.

2. Via a transparency and an overhead projector, demonstrate how the segmenting of the circle was done.

3. Show the circle graph below. Ask students to estimate what percent each segment is of the whole graph.

4. After students have made estimates, show them the actual percent and number of degrees in each segment.

Measurements
- A and B are each 90°, and each is 25% of the circle.
- C, D, and E are each 60°, and each is approximately 17% of the circle.

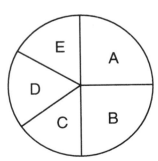

Name: _____ **Date:** _____

COMPARING QUANTITIES AND SIZES ON CIRCLE GRAPHS

Title: _____

_____

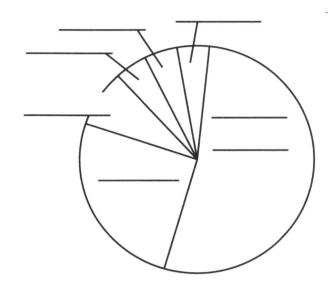

The circle graph above is about wheat production in the world's six greatest wheat-producing nations. The graph is divided into sections; each section represents a country. The total amount of wheat these countries produced is more than 434,000,000 metric tons. The size of each section tells how much of the total amount of wheat was produced by each country.

To Do:

1. The graph needs a title. Print this title on the lines above the graph: The World's Six Leading Wheat-Producing Countries.

2. Notice the blank spaces inside and outside the graph. Print the names of the countries in the blanks.

• Largest section: *United States*

• Second largest section: *China*

• Third largest section: *Brazil*

The remaining three sections show that three countries produced about the same amount of wheat. The countries are *Argentina, France, and Mexico.* Print their names in the remaining blanks.

The Gulf States Compared for Size

The states that border the Gulf of Mexico are Texas, Louisiana, Mississippi, Alabama, and Florida. The total area of these states is 487,000 square miles. The circle graph below will help you to see how the states compare in size.

To Do:

1. Texas occupies 55% of the land in the Gulf states. If you multiply .55 (55%) x 360° you will find that Texas will take up 198° of the circle. Write 198° on the line in the Texas section of the graph.

2. The four other Gulf states also occupy a certain number of degrees in the 360° circle.

Do the computation for each state by completing the multiplication problems below. Then, on the lines provided in the graph, write the number of degrees each state occupies.

$$\begin{array}{ll} 360° & 360° \\ \text{Florida: } \underline{x\ .14} & \text{Louisiana: } \underline{x\ .11} \end{array}$$

$$\begin{array}{ll} 360° & 360° \\ \text{Alabama: } \underline{x\ .10} & \text{Mississippi: } \underline{x\ .10} \end{array}$$

GULF STATES SIZE COMPARISON

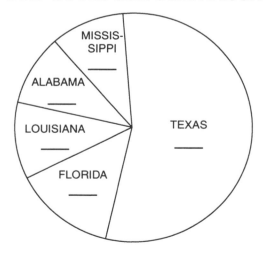

MAKING AND READING SQUARE GRAPHS AND SINGLE-BAR GRAPHS

Background (Square Graphs)

Square graphs are useful information sources. They are easy to read and understand at first sight, and they can provide detail with close examination. One significant advantage is that percentages can be determined quickly by counting blocks.

Following are some insights and procedures to follow in explaining square graphs.

1. The entire square represents 100% of the total entity.

2. There are 100 squares of the same size in the graph; therefore, each square is 1% of the whole. There should be ten rows of squares, each row containing ten squares.

3. The graph should be titled appropriately.

4. There should be a key to the graph. The key should tell what the various block sections represent.

5. The element taking up the largest portion of the graph should occupy the top squares, the next largest to follow, and so on.

The square in the next column may serve as a good example to show to students. A transparency would be especially helpful.

WORLD MOTOR VEHICLE PRODUCTION

E	E	E	E	E	E	E	E	E	E
E	E	E	E	E	E	E	E	E	E
E	E	E	E	E	E	E	E	E	E
E	E	E	US	US	US	US	US	US	US
US	US	US	US	US	US	US	US	US	US
US	US	US	US	US	US	US	J	J	J
J	J	J	J	J	J	J	J	J	J
J	J	J	J	J	J	J	O	O	O
O	O	O	O	O	O	O	O	O	O
O	O	O	O	O	C	C	C	C	C

Key: Each block represents 1%.

Europe $\boxed{\text{E}}$ United States $\boxed{\text{US}}$ Japan $\boxed{\text{J}}$

Other* $\boxed{\text{O}}$ Canada $\boxed{\text{C}}$

*Other includes countries such as Brazil, China, Italy, etc.

Background (Single-Bar Graphs)

A single-bar graph is an area graph. The entire bar represents 100% of what is being graphed. The bar may be divided into any number of segments up to 100. For example, a student's day was arranged on a single bar. The school day started at 9:00 AM and ended at 3:00 PM, for a total of 360 minutes. Eighteen divisions were made on the graph; each division represented 20 minutes. The graph below shows how the day was divided.

A STUDENT'S TYPICAL SCHOOL DAY OF SIX HOURS

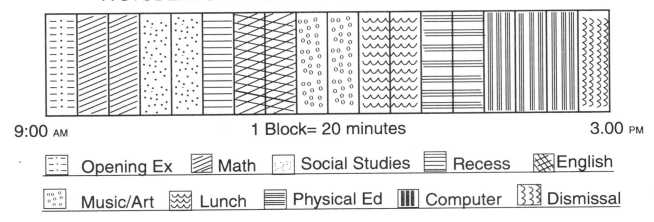

9:00 AM   1 Block= 20 minutes   3.00 PM

▦ Opening Ex   ▨ Math   ▦ Social Studies   ▤ Recess   ▨ English

▨ Music/Art   ▧ Lunch   ▤ Physical Ed   ▥ Computer   ▧ Dismissal

A SQUARE GRAPH COMPARES SOUTH AMERICA'S COUNTRIES

1. The total land area of South America is 6,888,935 square miles. What percent of the land area does each of the 13 countries of South America own? The square graph will tell you. Each block represents 1%. Count the blocks for each country, and print the amount in the table.

PERCENT OF LAND EACH COUNTRY IN SOUTH AMERICA OCCUPIES

BR	BR	BR	BR	BR	BR	BR	BR	BR	BR
BR	BR	BR	BR	BR	BR	BR	BR	BR	BR
BR	BR	BR	BR	BR	BR	BR	BR	BR	BR
BR	BR	BR	BR	BR	BR	BR	BR	BR	BR
BR	BR	BR	BR	BR	BR	BR	BR	AR	AR
AR	AR	AR	AR	AR	AR	AR	AR	AR	AR
AR	AR	AR	AR	PE	PE	PE	PE	PE	PE
PE	CO	CO	CO	CO	CO	CO	CO	BO	BO
BO	BO	BO	VE	VE	VE	VE	VE	CH	CH
CH	CH	PA	PA	EC	EC	GU	UR	SU	FG

Country	Percent of South America
Brazil	
Argentina	
Peru	
Colombia	
Bolivia	
Venezuela	
Chile	
Paraguay	
Ecuador	
Guyana	
Uruguay	
Suriname	
French Guiana*	

KEY

BR BRAZIL	AR ARGENTINA	PE PERU
CO COLOMBIA	BO BOLIVIA	VE VENEZUELA
CH CHILE	PA PARAGUAY	EC ECUADOR
GU GUYAMA	UR URUGUAY	SU SURINAME
FG FRENCH GUIANA		

* French Guiana is a colony of France.

2. How much larger, in percent, is Brazil than Argentina?

_____%; than Peru? _____%; than Paraguay? _____%; than Uruguay? _____%

3. Would the combined size of Colombia and Peru be greater than Argentina?

_____ Yes _____ No

4. If Brazil were to lose one-half of its territory, would it still be the largest country in South America?

_____ Yes _____ No

TWO SINGLE-BAR GRAPHS: THE GREAT LAKES AND FISHING

The two single-bar graphs on this page will help you learn how to read and use such graphs.

The Great Lakes of the United States

The graph that follows compares the sizes of the Great Lakes. Notice that the graph is divided into 50 spaces. Each space is 2% of the entire graph.

1. The Great Lakes cover a total area of 94,150 square miles. What percent of those square miles does each lake occupy? Refer to the graph below to complete the table.

Lake	Percent of the Whole
Superior	
Huron	
Michigan	
Erie	
Ontario	

SIZE COMPARISON OF THE GREAT LAKES

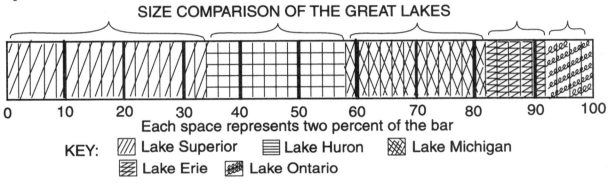

Each space represents two percent of the bar

KEY: ▨ Lake Superior   ▤ Lake Huron   ▩ Lake Michigan
      ▨ Lake Erie   ▨ Lake Ontario

The graph below can be completed to show the world's five leading fishing nations in a recent year. The total amount of fish the five nations caught was 53,359 metric tons. The table tells what percent of the total each nation caught.

To Do:

1. Find in the table the percent of fish caught by each country.

2. The country that is recorded first is the country with the greatest catch. The country with the second greatest catch is recorded next, and so on.

3. To record a country's percentage of total catch, look in the key to the graph to see what kind of symbol is used for that country. Remember that each space is equal to 2 percent of the whole fish catch.

Notice that China has been recorded to help you get started. China occupies 23 spaces for its 46%.

Country	%	Spaces	Country	%	Spaces
China	46	23	Japan	14	7
Peru	16	8	United States	10	5
Chile	14	7			

FISH-CATCH OF THE FIVE LEADING FISHING COUNTRIES

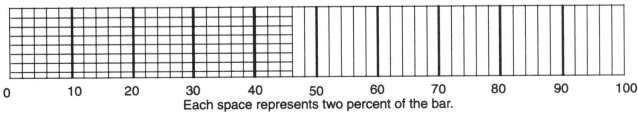

Each space represents two percent of the bar.

Key: ▤ China   ▨ Peru   ▨ Chile   ▩ Japan   ▥ United States

MAKING AND READING LINE GRAPHS

Background

Line graphs show the rise and/or fall of many different kinds of things such as a person's income, an industry's income, the stock market, SAT scores, and so on. A quick glance shows the direction and even the rapidity of the grafted element.

The information given in a paragraph in which time and quantities are stated can be confusing. The following paragraph is an example.

In September, 1994, the apple harvest in Spartan Valley was 4650 bushels. In 1995 the harvest was 5000 bushels, but in 1996 there was a drop in the number of bushels to 4200. However, there was a comeback in 1997, with a harvest of 5500 bushels. From that point on there was slow but steady gain in bushels harvested, with 6000 bushels in 1998 and 6300 bushels in 1999.

This same information could be easily shown and read on a line graph, as shown below.

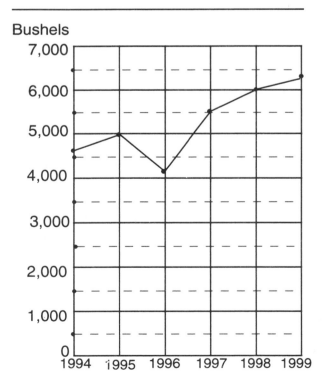

Propagandists and others who are interested in creating either positive or negative impressions will sometimes use line graphs for their purposes. All line graphs should be composed of squares. A line graph that is composed of rectangles can be quite deceptive.

If one wishes to convey rapid growth, the rectangles can be vertically arranged. Then, when the line is printed, the effect is of rapid and, sometimes, dramatic growth (Figure 1). If the rectangles are horizontally arranged, the indication is that there was slow growth (Figure 2). The first two line graphs that follow show how the same information may be distorted. The correct, and honest, graph composed of squares, is shown in Figure 3.

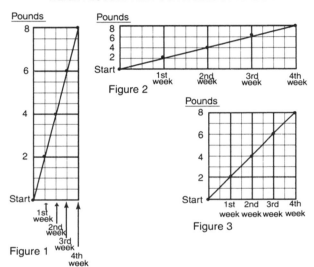

Suggestions for Teaching

The printed paragraph and the line graph in the first column can be made into a transparency and projected for your students to study. Ask them which presentation is easier to understand. Ask what other way the information could be presented that would also avoid the confusion of numbers in the printed paragraph. The answer is to show the information in a table.

The three graphs in the second column could also be shown via a transparency. Point out that the same figures are used in all three graphs, but the arrangement of the figures makes a considerable difference in the impressions given.

UNDERSTANDING TORNADOES

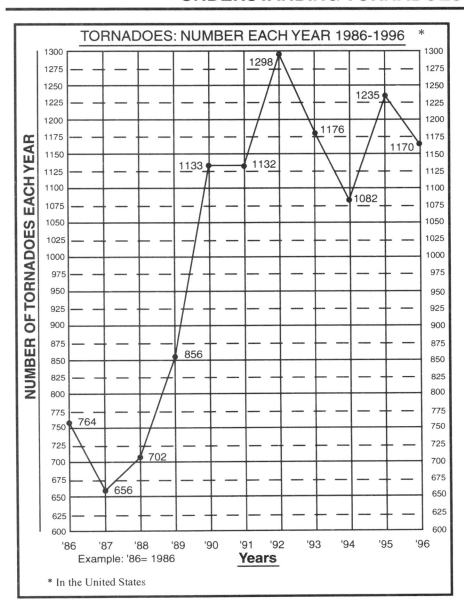

TORNADOES: NUMBER EACH YEAR 1986-1996 *

NUMBER OF TORNADOES EACH YEAR

1298
1235
1176
1170
1133   1132
1082
856
764
702
656

'86 '87 '88 '89 '90 '91 '92 '93 '94 '95 '96
Example: '86= 1986

Years

* In the United States

Tornadoes:

Tornadoes are violent, rotating columns of air that come down to Earth from clouds. Often a tornado resembles a funnel-shaped cloud; this is the reason they are often referred to as *twisters*. Sometimes, they look like tubes moving through the air.

Tornadoes move across the land at speeds up to 40 miles per hour, but they don't usually go very far, perhaps 15 or 20 miles.

As they twist, turn, and move over the land, they pick up loose objects such as wheelbarrows, carts, or even people. They are so powerful they can twist a roof from a house. The best protection from them is to go into an underground shelter or a basement.

To Do:

1. In what year did the least number of tornadoes occur? _____

2. In what year did the greatest number of tornadoes occur? _____

3. In what two years was the number of tornadoes almost the same? _____ and _____

4. In what two-year period did the number of tornadoes drop the most? _____

5. Another way to organize the tornado information is through the use of a *table*. With information gained from the line graph, print in the table the number of tornadoes that occurred each year. *Note*: Numbers next to the dots on the graph tell the actual number of tornadoes that occurred in that year.

Year	Tornadoes	Year	Tornadoes
1986		1992	
1987		1993	
1988		1994	
1989		1995	
1990		1996	
1991			

Name: _____ **Date:** _____

COMPLETING A LINE GRAPH OF POPULATION GROWTH

TITLE: _____

_____

_____

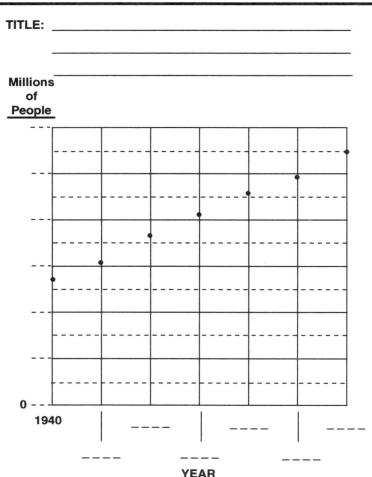

YEAR

The line graph above can be made to show the population growth of the United States from the year 1940 to the year 2000.

To Do:

1. Print the title of the graph on the lines above the graph:

POPULATION GROWTH OF THE
UNITED STATES FROM 1940 TO 2000

2. Print the years of population that the graph will show under the graph on the dashed (---) lines. The years in order are: 1940, 1950, 1960, 1970, 1980, 1990, and 2000.

3. The vertical axis of the graph has the heading: "Millions of People." Print the following next to its dashed lines from the bottom up (0 has already been noted): 50, 100, 150, 200, 250, and 300. Remember that each one of these numbers represents millions of people. For example, the 50 represents 50,000,000 people, or 50 million people.

4. Notice the dots on the vertical lines of the graph. The first dot on the 1940 line is between the 125 million and the 150 million line. The actual number is about 132 million.

Draw straight lines from dot-to-dot starting with the 1940 dot.

5. What was the population in 1940?

____ 100 million or ____ 132 million

6. What was the population in 1960?

____ 140 million or ____ 179 million

7. What was the population in 1990?

____ 220 million or ____ 249 million

8. The population in the year 2000 is an estimate. We won't know the exact population until another *census*, or counting, is made. What is the estimated population for the year 2000?

____ 255 million or ____ 275 million

SECTION 9

SOIL EROSION AND PREVENTION OF EROSION

9-1 All About Soil (Instructor) 104

9-2 What Happened to Riverside? I 105

9-3 What Happened to Riverside? II 106

9-4 Soil Erosion: Construction, Roadsides, and Strip-Mining I 107

9-5 Soil Erosion: Construction, Roadsides, and Strip-Mining II 108

9-6 Causes of Soil Erosion: Overgrazing, Bare Fields I 109

9-7 Causes of Soil Erosion: Overgrazing, Bare Fields II 110

9-8 The Importance of Sedimentation (Instructor) 111

9-9 Splash Erosion: Activity (Instructor) 112

9-10 Ways of Protecting Soil I 113

9-11 Ways of Protecting Soil II 114

9-12 Lesson Planning I: Developing Content, Skills, and Values (Instructor) 115

9-13 Lesson Planning II: Causes and Remedies for Soil Erosion (Instructor) 116

9-14 Lesson Planning III: Instructions to Students and Outcomes (Instructor) ... 117

9-15 Lesson Planning IV: Example Farm Design (Instructor) 118

ALL ABOUT SOIL

Background

It takes thousands of years for nature to form even one inch of topsoil. It takes so long because the chief ingredient of soil is rock. The rock must be reduced to powder. All of the following items tell something about how this is accomplished.

1. The force and friction of moving water wears away rock. The tiny pieces that are worn away then join water in its work. The sharp-edged little grains tumble and grate against larger rocks, gradually wearing them down.

2. When water in a rock crack or crevice freezes, it expands. The expanding ice will chip off pieces of the rock, or it may split the rock.

3. Water also breaks down rocks through a freezing-thawing cycle on rock surfaces. Each time there is a freeze followed by a thaw, you can be sure that a little more rock powder will be made.

4. Oxygen, carbon dioxide, and water are present in air. When they combine with the chemicals present in all rock, the rock tends to break down, or decay.

5. On a windy day have you ever been stung on your hands and face by dust particles? If so, you know it can hurt. This same pelting, grinding action of wind-driven dust can reduce great rocks, even mountains, to soil. If you think this is impossible, remember that nature has been "grinding away" at rocks for millions of years.

6. Earthworms are the world's champion soil-makers. They take already fine grains of soil and make them finer. The United States Soil Conser-vation Service writes in one of its publications, "The number of earthworms may range from a few hundred to more than a million per acre. Under favorable conditions between 200 and 1000 pounds of earthworms may be present in an acre of soil. The earthworms in an acre of soil pass several tons of soil through their bodies each year."

7. Considerable change of temperature within a few hours will help turn rocks into soil. During the day rocks are warmed by the sun. The heat causes the rocks to expand. After the sun goes down, the rocks cool. The cold causes the rocks to contract, or shrink. The stress of expanding and contracting acts on rocks in such a way that they may chip, split, or crack. People who have visited the desert regions of the world report that at night the splitting and cracking rocks sound like gunfire.

Making Soil Fertile

Powdered rock alone will not grow crops very well. It must be enriched with organic material. The organic material in topsoil comes mainly from decayed plants, the body wastes of animals, and animals that have died and decayed.

There are also living organisms in soil. These organisms can be divided into two main kinds: animal and plant. Most of the organisms are so small that a microscope is needed to see them. A pound of rich topsoil could contain thousands of organisms. These tiny living things are important because they make food for the larger above-the-ground plants from the decayed materials in the soil.

WHAT HAPPENED TO RIVERSIDE? I

1 Pioneers floating down the river on rafts chose Riverside as a good place to settle. The river would furnish transportation, water, and food. Cabins on the high bluff on the east side of the river would be easy to defend against Indian attacks. The low land to the north would be easy to farm. There were plenty of trees for fuel and lumber. The very life of the settlement would depend on the river.

2 Two hundred years later, the river and Riverside were still there. But, things had greatly changed.

3 Very few Riverside people went down to the river anymore. Parents warned their children not to swim there. They told the children that if they swallowed river water they would get sick. Maybe they would even die.

4 Nobody went fishing because there were no more fish except for a few "sunnies." Excursion boats used to take picnic parties downstream to Eagle's Point. But it had been thirty years since the last boat, the *River Queen*, had made the trip.

5 What had happened to the river? The answer is simple. The people had grown careless.

6 To save a few dollars the town had voted down the sewage treatment plant. So, the sewage from the town's cesspools and septic tanks gradually seeped into the river.

7 They had allowed the surrounding forests to be over-cut. No one planted seedlings to replace the cut trees. There was no cover to break the force of the raindrops when they hit the earth. There were few plants with roots to hold the soil particles. So, the soil washed into the river during heavy rains. The river became brown and dirty. The channel filled with sediment.

8 People dumped old cars, furniture, tin cans and other garbage over the edge of Indian Cliff. Some thought it was great fun to skim old plastic phonograph records from the top the cliff into the water. Rubber tires were rolled off the top of the cliff. Chemicals from the rusting, rotting garbage combined with the small plant life in the river. This used up most of the oxygen in the water. Of course, the fish suffocated, and the larger plant life died.

9 Industry had settled on the bank of the river north of town. There were no regulations for the treatment of waste from the factory. Thousands of gallons of industrial waste poured into the river every day. Smells and fumes that arose from the river were well known by people for miles around.

To Do:

1. Briefly list four reasons why early settlers chose to settle at Riverside.

a. _____

b. _____

c. _____

d. _____

2. Study the picture. Tell how industry polluted the air of Riverside.

_____

_____

_____

_____

_____

3. What is the number of the paragraph that tells why sewage from the town was allowed to pour into the river? _____

4. In what paragraph does the following sentence most naturally belong? *Nobody wanted to sail on a river that smelled and on which garbage floated.*

_____

5. Notice the tree stumps on the left side of the picture. Circle the three sentences in the story that tell why the rain that fell loosened the soil and led to erosion.

6. Circle the two sentences in the story that tell why the runoff from the dump killed fish in the river.

WHAT HAPPENED TO RIVERSIDE? II

7.   On the lines that follow tell what kind of pollution is taking place at each of the numbered places in the picture.

1 _____

2 _____

3 _____

4 _____

8.   Place A shows a sandbar building on the river as a result of soil erosion from the river bank. Print *sandbar* on the line. How will the sandbar affect boat travel on the river?

_____

_____

_____

SOIL EROSION: CONSTRUCTION, ROADSIDES, AND STRIP-MINING I

Soil erosion—the washing and blowing away of soil—results from several causes. Three of those causes are explained and illustrated on these two pages. Other causes of soil erosion will be explained on other pages.

After you complete these two pages and the other pages that follow you will be much more aware of soil erosion and its causes. For example, there may be new home construction in your neighborhood or your town. Study the construction site carefully for signs of erosion. Also, when riding in an automobile, look carefully at the sides of the roads. Are there road shoulders that have been left uncovered? Are there high embankments that are bare? If so, you know that there is a problem, and, if the problem is left uncorrected, one thing is certain: the erosion will continue, and the washed away soil will cause still more problems.

Construction

When houses are built, it is necessary to strip much of the land's vegetation before construction can begin. For as long as the ground remains bare, soil erosion will take place.

Although it is impossible to completely eliminate soil wash-away at construction sites, much can be done to reduce soil loss. Three helpful practices follow: (1) Do not strip the vegetation from the earth until just before construction is to begin. (2) As soon as possible after excavation, replant the land. (3) During construction, place a barrier around the construction site. Soil runoff will be stopped at the barrier.

To Do:

1. On line (1) beside picture A, write a brief caption for the picture.

2. At (2) write at least one suggestion of your own that would help reduce erosion at construction sites.

3. Where will most of the loose soil shown in the picture eventually be deposited? Answer at (3).

4. Study the construction site carefully. Would you say that this construction was recently started? Explain your answer at (4).

Roadsides

Roadside erosion takes place when banks and ditches that border roads are left without protective cover. During heavy rains, great loads of sediment are carried in the water runoff from the road. Eventually, the sediment will clog sewers and streams. Roadside ditches should be planted with grass or some other soil-holding plant. Sometimes it is useful to pave a ditch with asphalt, stones, or bricks. Road banks can be protected by retaining walls. Also, banks can be graded and sloped and then planted with grass or vines.

To Do:

1. At (1) beside picture B, write a brief caption for the picture.

2. At (2) list two ways that a ditch can be protected from soil erosion.

3. At (3) tell how a road bank can be protected.

4. A little more erosion, and the guard rail on the side of the road could be displaced. Why would that be dangerous? Briefly explain at (4).

Strip-Mining

Taking minerals such as sand, stone, and iron ore from the surface of the earth is called strip-mining. Giant bulldozers and mechanical shovels dig out the minerals and load them on conveyors or trucks for further processing.

Strip-mining contributes to soil erosion because the earth loses its protective covering of plants and trees. The constant maneuvering of the heavy machines further loosens the soil, which then is carried away by water and wind.

To Do:

1. At (1) beside picture C write a brief caption for the picture.

2. How does strip-mining contribute to soil erosion? Write you answer at (2).

3. Suppose you could make a law regulating erosion at strip-mine sites. What law would you make? Write your answer at (3).

Name: _____ **Date:** _____

SOIL EROSION: CONSTRUCTION, ROADSIDES, AND STRIP-MINING II

A

(1) _____

_____

(2) _____

_____

_____

(3) _____

_____

(4) _____

_____

B

(1) _____

_____

(2) _____

_____

_____

(3) _____

_____

_____

(4) _____

_____

C

(1) _____

_____

(2) _____

_____

_____

(3) _____

_____

Name: _____ Date: _____

CAUSES OF SOIL EROSION: OVERGRAZING, BARE FIELDS I

Overgrazing

1 Cows, sheep and horses all have hooves. When a few of these animals graze on good pasture, little damage occurs to the grass and sod. But when the hooves of many animals cut, twist, and turn in a small area, the result is loosened soil.

2 Too many animals in an area also means that there will not be enough food to satisfy their hunger. So, each blade of grass is eaten down to its very roots. In this way the protective covering of the soil is destroyed.

3 The loosening of soil and the removal of protective covering by animals is called *overgrazing*. When the rains fall, the snows melt, and the winds blow, there will be nothing to prevent the soil from floating and blowing away as a result of overgrazing.

The picture at the bottom of the first column shows one possible result of overgrazing. The cut was begun when a mere trickle of water began to find its way over bare land. In time, the water gouged an ever-deepening and ever-widening trench. The gully, as this kind of cut in the earth is called, will get even larger if nothing is done to stop run-off waters from racing through it.

Bare Fields

Land that is left bare for even a short period of time will probably begin to erode, or wear away. Very often, the first step in the erosion process is for soil particles to float off the field in "sheets" of water. This kind of wearing away is called *sheet erosion*.

But soon the water begins to run through the soil in *rills*, or very small streams. When soil is carried away in this manner, *rill erosion* is taking place. It is possible for sheet erosion and rill erosion to be working the same field at the same time. These two forms of soil wear-away are shown in the picture at the bottom of the second column.

If rains continue to fall on the eroding field, the rills will probably become longer and deeper. *Gullies* will form. The picture at the bottom of the first column shows an advanced case of *gully erosion* that started as a rill.

Overgrazing could have led to the formation of this deep gully.

CAUSES OF SOIL EROSION: OVERGRAZING, BARE FIELDS II

To Do:

Overgrazing

Paragraph 1

What three words tell the actions of animals' hooves that result in soil becoming loose?

_____ _____ _____

Paragraph 2

What happens to grass when too many animals graze in a small area?

_____

_____

Paragraph 3

1. What three things related to weather help cause erosion?

_____ _____ _____

2. What paragraph best fits the title listed below? Check the best answer.

Weather Can Be a Contributor to Soil Erosion

___Paragraph 1   ___Paragraph 2   ___Paragraph 3

3. Circle the two sentences that tell how gullies are formed.

Bare Fields

1. Circle the sentence that tells how sheet erosion occurs.

2. Explain how rills become gullies.

_____

_____

_____

3. The drawing below shows a bare field behind the cattle and the fence. Draw a number of rills across the field.

Learning from Pictures

Picture 1

1. Notice the man standing in the gully in the picture. Compare his height with the depth of the gully.

_____

_____

2. What is there about the gully that leads you to think that it has been a long time in forming?

_____

_____

3. Suppose that it is your job to do something to help stop erosion of the gully. What will you do?

_____

_____

_____

4. Imagine the man at the bottom of the gully is talking about the gully. What do you think he might be saying about the gully?

_____

_____

_____

Picture 2

1. Notice the blank lines below the picture. On the lines write your own caption (one sentence) for the picture.

2. In what way are the cattle in the picture contributing to erosion?

_____

_____

3. What do you think will happen to the rills in the field during the winter months?

_____

_____

_____

4. A line of trees across the top of the bare field would be helpful in slowing down wind erosion. Draw a line of trees across the top.

THE IMPORTANCE OF SEDIMENTATION

Suggestions for Teaching

1.   Notice the three bottles pictured on this page. Bottle #1 shows water and sediment typical of water from a stream running through pasture land; #2 shows water from a woodland stream; #3 shows water from a cultivated field that has not been well managed. It is easy to tell which bottle contains the least sediment—the bottle from the pasture stream. If desired, these facts can be verified by obtaining water from streams as described just after a heavy rainfall.

2.   On the board draw three bottles, but omit the "dotting." Then, with information gained from the above paragraph show the degree of sedimentation in each bottle and explain the origins of the samples. Label the bottles as follows: #1, Pasture; #2, Woodland; #3, Cultivated Field.

3.   Orally read the article that follows. As you read, make a brief outline on the board of the article. (Students copy the outline.)

Why Sediment Washed from Farmland Is Troublesome

1.   Topsoil erosion reduces farm productivity.
2.   Water-supply reservoir capacities are reduced. There is less water for:
a.   Drinking*   c.   Irrigation
b.   Recreation*   d.   Hydroelectric power
3.   Road and railroad ditches become filled; culverts plugged; channels filled.
4.   Harbors have to be dredged.
5.   Floods are more severe.
6.   Fish are suffocated, eggs covered.

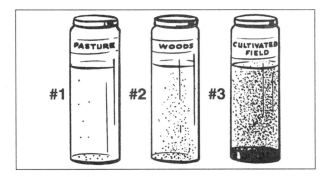

© 2000 by The Center for Applied Research in Education

Why is soil erosion so devastating? The information that follows, distributed by the United States Soil Conservation Service, helps us to know the answer to this question.

There is an important story in the three bottles—the story of how sediment washed from farmland hurts the farmer and city dweller in many ways.

Sediment carried by streams hurts the farmer first because it is part of his farm that is being carried away. Much of the sediment is topsoil. Topsoil takes thousands of years to form, so it cannot be easily replaced.

After the sediment leaves the farm, some of it gets into streams and begins to affect everyone. More than 3200 water-supply reservoirs are losing water-storage capacity each year to sediment. Water bills are higher because the water must be filtered.

Seventeen percent of the electric power generated in the United States comes from hydro-electric plants. The storage reservoirs serving these plants are gradually filling with sediment that displaces the water the plants need.

Sediment fills road and railroad ditches, plugs culverts, and clogs stream channels. Clearing these water passages and/or raising the bridges that cross them costs time and money. All this increases taxes. The national sediment damage amounts to millions of dollars annually.

Floods are more frequent and more serious, partly because the streams become choked with sediment and, thus, incapable of carrying flood-waters.

Silt harms fish by covering up their spawning grounds and shading out light necessary for food growth. Many fish actually die during floods when their gills are clogged with silt.

Soil and water conservation measures applied to farm and ranch land will greatly reduce sediment. Erosion that causes sediment deposition can be reduced up to 90 percent with soil and water conservation measures. Growing grass and trees will reduce erosion greatly. This is true because they give protective cover and add organic matter, which helps the soil take in water more readily.

---

*Optional—not actually mentioned in the article.

SPLASH EROSION: ACTIVITY

Background: The United States Soil Conservation Service has made studies that show that between one and one hundred tons of soil per acre may be splashed up on bare soil during a rainstorm. The force of rain dropping breaks the soil clods into small particles that are easily carried away by running water. The particles, called sediment, eventually find their way into streams and rivers and are then deposited in channels or large bodies of water such as lakes, reservoirs, and oceans. Of course, the fields from which the soil came gradually become less and less productive.

If rain falls on land that is covered by crops, grass, weeds, bushes, or trees, the force of the falling rain is lessened because it first strikes leaves and stalks. The water then drips slowly and soaks into the earth gradually. Under these conditions, the soil loss is relatively light and water is able to be stored for use during times of little rain.

Experiment Purpose: To demonstrate the effects of falling water on covered and uncovered soil as a simulation of what actually happens to a field during a rainstorm

Materials: ❏ Two wide-mouthed jars with lids that have been perforated by an ice-pick or nails, etc.
❏ Two sprinkling cans filled with equal amounts of water
❏ Large foil-type roasting pan such as is used for roasting turkeys

Procedure: 1. Place the jars about 3" apart in the roasting pan.
2. Place the lids upside down on top of the jars.
3. Fill both lids with soil. Cover the soil in one jar lid with grass clippings; leave the other soil uncovered.
4. Sprinkle the soil in the lids with water from a height of about three feet, making sure that each jar lid receives equal amounts of water.
5. To observe:
❏ The amount of soil splashed from each lid as seen in the pan. Note color.
❏ The amount, color, and quantity of the water in each jar after sprinkling.
❏ The time it takes for the water to soak through each lid.

Drawings:

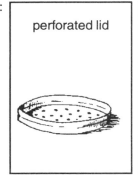

perforated lid

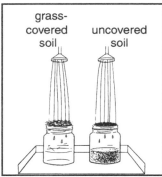

grass-covered soil   uncovered soil

In fields signs of splash erosion can be observed. The coin is on a pedestal after beating rain carried away the soil.

Conclusions: 1. The grass-covered soil splashed less, and the water splashed was clear.

2. The grass-covered soil's jar had clearer water within it.

3. It took a longer time for the water to filter through the grass-covered soil.

4. Overall conclusion: Fields with cover will withstand erosion better than fields without cover.

Question: Often, when grass seed is planted, the seed is mixed with a sticky, thick substance, and then sprayed on the soil. The substance then hardens. What is the reason for seeding lawns this way? (The substance hardens after being sprayed and acts as cover, thereby preventing splash erosion and the subsequent washing away of seeds and soil.)

WAYS OF PROTECTING SOIL I

Contour Farming

Contour farming is plowing, planting, cultivating, and harvesting across a slope of land rather than down the slope. Contour strip cropping is the practice of alternating strips of row crops such as corn, with strips of cover crops, such as clover.

There are several advantages to contour farming. Contours hold water long enough for it to soak into the ground. Also, the speed of the water that does run downhill is reduced because the contours act as catch basins. The slower that water moves, the less soil it will cut away and carry.

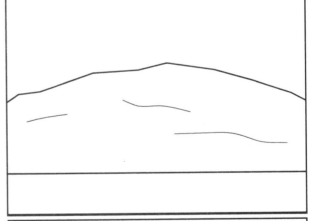

Terracing

In terracing, a series of "steps" reduces the unbroken length of a slope. The level steps slow the flow of water down a slope. This reduces soil loss.

Sometimes terraces are constructed in such a way that collected water is guided off the slope into ponds. The water in the ponds can then be used for irrigation, fishing, or recreation.

Protective Covering

Keeping land covered with close-growing grasses, shrubs, and trees is called protective covering. The branches and leaves of the trees reduce splash erosion by breaking the fall of raindrops before they reach the ground. The roots of the cover plants help hold soil in place. The litter from the plants—leaves, seeds, dead branches, bark—absorbs and holds water.

Protective covering can be used in any situation where bare ground is exposed: road cuts, sand dunes, gullies, etc. **To Do:** *Show that you understand protective covering by drawing grass, small brush, vines, and branches on the bare hillside in the picture.*

Willow Planting

Willow planting prevents the washing away of soil from the banks of streams. The tough, stringy roots of the willows lace together. This network of roots helps hold the soil firm.

Willow plantings are especially useful where rivers curve. The cutting action of rivers is greatest on downstream, outside curves.

WAYS OF PROTECTING SOIL II

Wind Breaking

If the winds that sweep across bare land are not interrupted, they will pick up soil particles. Bit by bit, soil, in the form of dust, is removed from productive farm land. In this way, million of tons of topsoil have been blown off the Great Plains and other parts of the United States.

One way to fight wind erosion is through the use of *wind breaks*. A wind break consists of a line of trees or hedges planted across the paths of the prevailing winds of a region. The wind breaks act as barriers so that the forces and dust-carrying power of the winds are greatly reduced.

To Do:

1. Circle the sentence in the story that does the following:

a. gives a definition of contour farming.
b. gives a definition of strip cropping.
c. tells how terraces slow down water.
d. tells where willow planting is most effective.
e. gives a definition of wind breaks.

2. In what two ways are contours useful?

_____

_____

3. Why are willow trees used especially on the curves of rivers?

_____

4. How do bushes, branches, and leaves lessen soil erosion caused by falling rain?

_____

_____

_____

_____

5. On the blank lines in each picture write a suitable caption for the picture.

6. The drawing below shows the five ways of protecting soil. At each numbered place in the picture write the term that best describes what is being shown: Contouring, Terracing, Willow Planting, Wind Breaking, Protective Covering.

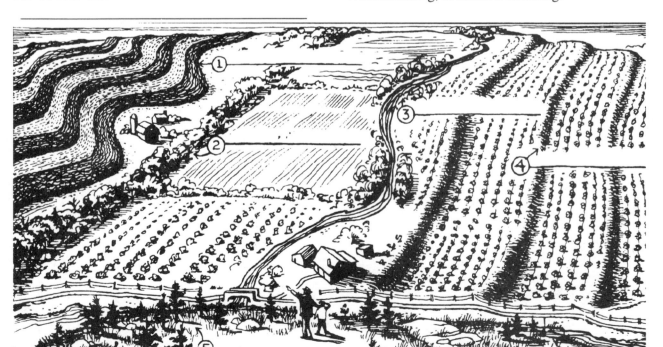

LESSON PLANNING I: DEVELOPING CONTENT, SKILLS, AND VALUES

An effective lesson plan has several important parts, as follows:

1. Objectives

a. Understandings

An understanding is a generalization or conclusion that results from combining the details of a lesson. In other words, it helps students to see the "forest" and not just the "trees."

b. Skills

Every lesson should result not only in subject matter acquisition, but also in skill development. When a lesson is taught with emphasis on skills as a means of gaining the subject matter, both elements of learning are enhanced.

c. Values

Values are an important teaching consideration. If what is learned does not have a positive effect on group or personal thinking and action, something significant has been lost. For example, if a lesson on water conservation does not result in conscious efforts to reduce water use, especially in times of drought, what was the purpose of the lesson other than an intellectual understanding?

2. Introduction

The introduction of a daily lesson plan should have three basic elements.

a. Each lesson should begin with a brief review of the previous lesson. A review is an aid to retention. There is much truth in the expression, "We remember what we rehearse."

The review may be a set of brief oral or written questions to which students react in writing. Another approach might be to have a number of students recall a fact or detail of the previous lesson. After ten or more students have expressed a learning, other students who are listening will have their recall refreshed.

b. A brief interest-arousing element regarding the upcoming lesson helps students develop a mind-set related to what will be taught. For example, if the upcoming lesson is about water conservation, a newspaper article might be read that reports decreased water levels in reservoirs and wells.

c. The basic problem of the new lesson can be stated in the form of a question. The question should encompass the details of the lesson. The answer to the question should be contained in the understanding objective. A typical question might be, "What can be done to conserve the use of water in our homes and community?"

3. Development

The development of a lesson consists of a step-by-step march through it. The steps should be listed in chronological order and should indicate what the teacher and students are going to do. The steps should be brief, but not so brief as to omit important details. Whenever possible some kind of activity should take place.

4. Summary

The conclusion of the lesson should be a brief review of its basic elements. The understanding objective of the lesson should be expressed by the students with the help of the instructor. It would be helpful to write the conclusion on the chalkboard. If students have notebooks it would be well for them to write the conclusion and some of the supporting details.

The teacher should mention the skills that have been developed through the lesson. It is important that students know what it is that they are becoming better at doing.

Finally, a brief discussion should take place relative to the value objectives. If this is done, the chances that the children will act upon values will be increased.

Additional Information

The lesson plan on the next page is an example of a "Presentation Activity" lesson. In such a lesson the instructor leads the class for a portion of the lesson, perhaps 30% of the total time. In the remaining time students are actively involved in making their conservation models.

It should be noted that this lesson takes two periods to complete. Also, a set of instructions is provided for completing the model farm.

Also note that the second page following this one contains a list of skills and instructions for completing the model farm. The third page following is an example of a farm design.

LESSON PLANNING II: CAUSES AND REMEDIES FOR SOIL EROSION

I. Objectives

Understandings

The single and combined actions of gravity, wind, and water cause soil to erode, but the erosion process can be controlled through certain preventative measures.

Skills

• Planning   • Following directions
• Translating a two-dimensional graphic representation into a three-dimensional model
• Sharing and working cooperatively with others

Values

To develop a positive attitude toward and concern for the protection of the environment

II. Procedures

Introduction

1. Review lesson on watersheds. Allow several minutes for study of notes. Project transparency of a watershed. Ask for volunteers to explain the "Journey of a Raindrop."
2. Tell a story of a farm family forced to leave their home due to soil depletion.
3. State problem of the lesson: How is soil eroded? How can erosion be prevented?

Development

1. Explain how strip mining leads to erosion when a stripped area is left uncovered.
2. Explain how construction sites and uncovered road embankments contribute to soil erosion. Construction sites are left uncovered for long periods of time. Steep embankments left uncovered allow significant runoff of soil.
3. Explain how overgrazing leads to gully erosion and how bare fields lead to sheet and rill (rivulets of water carry off soil) erosion. Animals eat grass down to the roots, thus leaving the soil without cover.
4. Blow sand across a flat pan to demonstrate how wind erodes topsoil.
5. Discuss, illustrate, and demonstrate ways of conserving soil.
   a. Contour plowing: On a lump of clay resembling a hill, draw vertical rows in which water flows freely and carries soil. Next, eliminate vertical rows and replace with horizontal rows that hold water back and allow moisture to drain into the soil.
   b. Terracing: On the chalkboard draw steps (terraces) with slight ridges on the outside edge of the step to slow water runoff. Terraces may also be illustrated on the clay model.
   c. Protective covering: Bushes, grass, shrubs, etc. can be placed over stony areas to prevent erosion.
   d. Willow planting: Willow trees that have intricate root structures can be planted on downstream, outside curves of rivers and streams; the roots hold soil.
   e. Wind breaks: Rows of trees can be planted to break the force of prevailing winds. Demonstrate by placing twigs to represent trees on the edge of a flat pan. Distribute sand in the pan. Blow against the trees, and point out the minimum of displaced sand.
   f. Strip cropping: Strip cropping is the practice of planting alternate rows of crops on the same field. This practice helps conserve moisture and soil. For example: clover alternates with corn; the clover conserves water, acts as cover, and replenishes soil with nutrients.
6. Activity: Constructing a 3-D conservation farm. Distribute and explain instruction for making the model. See attached.

Summary

1. Various students will show their models accompanied by explanations.
2. Solicit a generalization (and write on board) relative to soil erosion and conservation consonant with objective #1.
3. Ask: What did we learn to do better? (Make a diagram and follow it.)
4. Ask: What can you/we do to decrease soil erosion?

III. Teaching Aids

1. Clay shaped into a mound
2. Flat plate for wind erosion and wind break demonstrations
3. Twigs, sand
4. Styrofoam boards (8" × 10") for model farm

LESSON PLANNING III: INSTRUCTIONS TO STUDENTS AND OUTCOMES

How to Make Your Model Conservation Farm

1. On a piece of paper about the same size as your styrofoam board, sketch a plan of your farm. Show how you will divide the farm's surface into fields, river, farmhouse, etc. Your sketch should show the following:

   - terraces
   - strip cropping
   - protective covering
   - contours
   - wind breaks
   - a farmhouse
   - willow planting on a river curve (Show the direction of the current with arrows.)
   - anything else that will help your farm look real—barns, windmills, animals, people, silos, fences, flowers, telephone poles, etc.

2. Use your sketch to make a three-dimensional model. Some things to keep in mind:

   - Use as little glue as possible—a little goes a long way, dries faster, and looks better.
   - Use crayons or colored pencils for coloring.
   - Punch holes in the styrofoam to help your trees, etc., stand up. Drop a small amount of glue into the holes to hold the trees fast.
   - Cardboard can be used to make terraces.
   - Try not to make your model too crowded.

3. Work neatly. Keep your area clean. Protect your desk with newspapers.

4. Work quietly. You may talk with others and move around, but remember that you have to live up to this privilege by using good judgement.

Conservation Lesson: Some Outcomes for the Students

Skills and Processes

- Planning, organizing
- Estimating (scale and distance)
- Hand dexterity
- Interpreting pictures
- Inferencing
- Translating abstract symbols into reality via a model
- Arriving at a conclusion after considering data
- Following directions

Subject Matter

(See pp. 107–115)

Values

- Learning conservation of materials
- Sharing
- Cooperating
- Assuming responsibility for one's conduct
- Deriving satisfaction from hard work
- Developing possible hobbies
- Developing empathetic feelings for the less fortunate
- Developing sympathy for the physical environment; a protective attitude toward the environment
- Developing appreciation for the value of soil
- Being creative within structure
- Deriving satisfaction through the creation of a tangible object

LESSON PLANNING IV: EXAMPLE FARM DESIGN

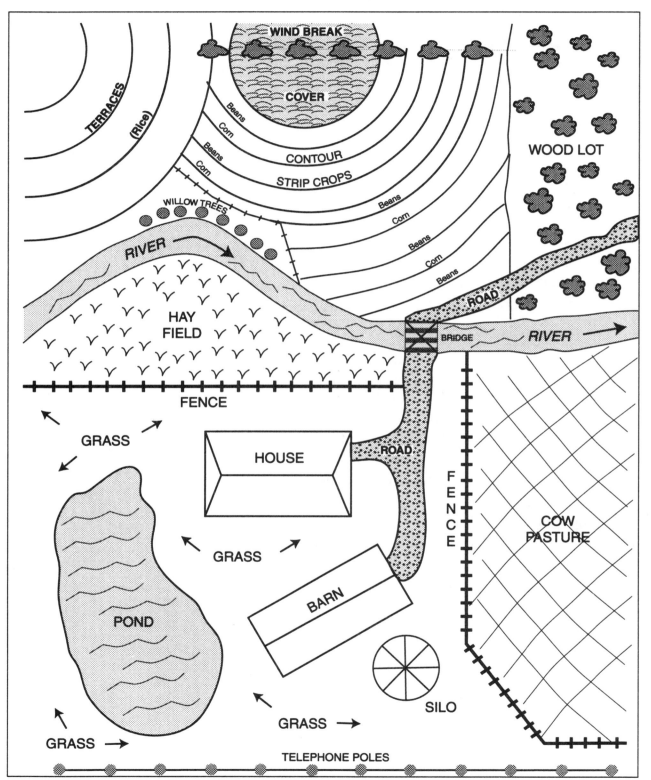

WATER: ITS SOURCES AND ITS POLLUTION

10-1 All About Water (Instructor) 120

10-2 "Water" Is an Important Word (Instructor) 121

10-3 Water: Too Precious to Waste (Instructor) 122

10-4 Understanding Watersheds 123

10-5 More About Watersheds I (Instructor) 124

10-6 More About Watersheds II 125

10-7 The Water Cycle: An Explanation (Instructor) 126

10-8 Understanding the Water Cycle 127

10-9 Water from Underground Sources I (Instructor) 128

10-10 Water from Underground Sources II 129

10-11 Sources of Fresh Water: Surface I 130

10-12 Sources of Fresh Water: Surface II 131

10-13 Water Pollution: Spilled Oil I (Instructor) 132

10-14 Water Pollution: Spilled Oil II 133

10-15 Oil Spill in the Shetlands: An Environmental Disaster I 134

10-16 Oil Spill in the Shetlands: An Environmental Disaster II 135

10-17 Thermal Pollution: What Is It? How Can It Be Controlled? 136

ALL ABOUT WATER

• Water that has been polluted can be returned to its original condition. When humans purify water they use the same process that nature uses, except that they speed up the process.

• There are several ways to remove salt from water. One of the most common methods is distillation. In this process, salt water is boiled so that the water separates from the salt and becomes steam. When the steam cools, it returns to its water form minus the salt.

• The weight of water varies slightly with its temperature. However, at 70°F, water weighs slightly more than 62 pounds per cubic foot.

• Lakes, streams, and ponds occupy about 2% of the area of the United States.

• An advantage of using water power to make electricity is that the water is returned unharmed to rivers and streams.

• A person could not live more than eleven days without water.

• The amount of water available on earth today is just about what it has always been. But, because the earth's population has greatly increased, the same amount of water must serve far more people.

• On its way to the sea, water picks up from the earth many different minerals, including salt. The salt is deposited in the oceans. This process has been going on for millions of years. Thus, the sea becomes saltier each year.

• Pure water is virtually colorless. However, water may appear blue, green, or brown. Reflection of the sun's rays, reflection of the sky, and the presence of soil particles all affect the appearance of water.

• Water covers about three-quarters of the earth's surface.

• The amount of water used by different manufacturers to make the same product varies. For example, to make one ton of steel, from 1400 to 6000 gallons of water may be used.

• In the United States, more water is used for agriculture than for all other uses combined.

• In 1900, the average person living in a town or city used about 95 gallons of water a day for all purposes. But, in today's world of washing machines, backyard swimming pools, etc., the average city dweller uses over 150 gallons of water a day.

• The water an average person drinks plus the water he gets from other foods adds up to about one ton of water a year.

• Most of the water we use comes from surface storage places. However, there is about six times as much water below the earth's surface as there is on its surface.

• When water freezes, it takes up more space. This is why pipes sometimes burst when the water in them freezes.

• Disease germs such as typhoid and cholera can live in water. To kill such germs, chlorine gas is sometimes added to the water.

• About 65% of the weight of the human body is water. Blood is mostly water, and muscles are more than 80% water.

• The average rainfall in the United States is about 30" per year. Some places on the planet receive less than 5" of rain yearly, while other places get as much as 120" of rain.

• If water contains an excessive amount of minerals, such as lime, it is called "hard water."

• At 32°F and above, the molecules in water are very active. A drop in temperature causes the molecules to slow down. Water becomes ice when the temperature falls below 32°F.

• A steel needle will float on water. Why is this so even though the needle is several times more dense than water? The reason is that the water has "surface tension." The molecules on the surface of the water cling together and form a kind of cushion. The needle sinks slightly, but remains supported. Disturb the surface tension, and the needle will sink.

"WATER" IS AN IMPORTANT WORD

Water is so common that many people take it for granted; millions of gallons are wasted each year. It's not until a drought occurs that the importance of water is fully realized. The following information and activity should impress students with water's value.

Literature
"Do not let your chances like sunbeams pass you by,
For you never miss the water 'till the well runs dry."
—Rowland Howard

"Water, water everywhere,
and all the boards did shrink;
Water, water everywhere,
Nor any drop to drink."
—Samuel T. Coleridge
(in reference to being adrift on the salt water ocean.)

"Still water runs deep."

(Author unknown; refers to the thought that people who do not often speak may still have profound thoughts.)

"Little drops of water, little grains of sand,
make the mighty ocean and the pleasant land."
—Julia Carney

Verb Synonyms (to water)
irrigate, moisten, drench, wet, soak, steep, sprinkle, dampen, hose, spray, douse, slosh, inundate, flood

Noun Synonyms (water)
H_2O, rain, dew, condensation, moisture, tears, perspiration

Adjectives (related to water)
watery, soggy, sopping, sopping wet, damp, dank, moist, soaked, drenched, saturated, waterlogged, boggy, swampy, quaggy, miry, marshy, mushy, spongy, squashy, squishy

Water Words Activity
Place the students in groups of two or three. Ask them to list terms that in some way use the word *water*. Tell students that in making their lists they are not to use the dictionary. Give an example from the list below. After a short time, perhaps ten minutes, ask each group how many terms they have listed; then have the terms read orally. It may be helpful to make a complete list of all the listings. *Note*: *Webster's New Collegiate Dictionary* lists more than 125 separate and distinct entries with *water* as the first word in a term.

Water Words

1. water bag	9. waterfront	17. water meter	25. water rat
2. water ballet	10. water gap	18. water mill	26. watershed
3. water bed	11. water heater	19. water moccasin	27. water ski
4. water bird	12. water hole	20. water pipe	28. water snake
5. water boy	13. water level	21. water pistol	29. water spot
6. water bug	14. water lily	22. water polo	30. waterwheel
7. watercolor	15. waterline	23. water power	31. water wings
8. waterfall	16. watermelon	24. water proof	32. waterworks

WATER: TOO PRECIOUS TO WASTE

Benjamin Franklin, who had the knack of expressing profound thoughts in simple language, wrote in *Poor Richard's Almanac*, "When the well's dry, we know the worth of water." The fact is that with the world's population increasing—doubled from 2½ billion to 5 billion from 1950 to 1990—and with uses for water increasing, many wells have run dry and many new reservoirs are needed. So, it behooves us to think about ways to conserve water.

Suggestions for Teaching

1. Discuss the thoughts contained in the above paragraph.

2. Present and discuss the following approaches to conserving water both inside and outside of homes.

3. Follow up with your students' ideas for conserving water. The symbols in the boxes below will help stimulate their thinking.

Water-Saving Ideas

☐ Store chilled water in the refrigerator to eliminate the need to run water into the sink until it gets cold.

☐ Water lawns in the cool of the evening rather than in the hot sun. In this way water will sink into the ground rather than evaporate.

☐ Keep a cover on outside swimming pools when they are not in use—to decrease water loss from evaporation.

☐ Collect water from leaders on the house so the runoff can be used for other purposes.

☐ Place large blocks in the water tanks of toilets. This may displace as much as one gallon of water and thus save a gallon in every flush. In a family of four this could amount to 12–15 saved gallons each day.

☐ Test for a leaking toilet by adding food coloring to the tank—within 30 minutes color will appear in the bowl if there is a leak. Do not rely on the sound of water leaking because the sound may be undetectable by the human ear.

☐ While waiting for the water to turn warm before taking a shower, catch the water in a bucket. Use it later for watering plants and other uses.

Water-Saving Ideas

The symbol in each box should lead to suggestions for saving water.

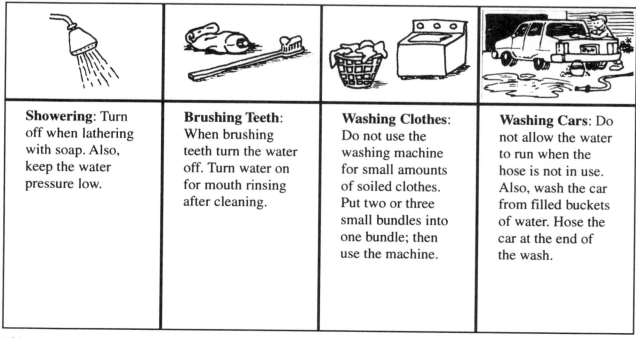

Showering: Turn off when lathering with soap. Also, keep the water pressure low.	**Brushing Teeth**: When brushing teeth turn the water off. Turn water on for mouth rinsing after cleaning.	**Washing Clothes**: Do not use the washing machine for small amounts of soiled clothes. Put two or three small bundles into one bundle; then use the machine.	**Washing Cars**: Do not allow the water to run when the hose is not in use. Also, wash the car from filled buckets of water. Hose the car at the end of the wash.

Name: _____ **Date:** _____

UNDERSTANDING WATERSHEDS

A watershed is an area surrounded by high land where all the rivers and streams drain into a main river or directly into a large body of water such as an ocean. On the map above, all the land and rivers within the dashed lines are part of the *Arapaho Watershed*.

To Do:

1. Write the title of the watershed in the box in the bottom right corner of the map.

2. With a heavy stroke, connect the dashed lines that mark the boundaries of the Arapaho Watershed.

3. What are the three main rivers that flow into the Arapaho River?

_____

_____

_____

4. The *source* of a river is where it begins. Print SOURCE at the beginnings of the four labeled rivers, where they are already marked with an S.

5. The *mouth* of a river is where it empties into another larger river or a large body of water. Print MOUTH at the ends of the various large and small rivers marked by an M.

6. A *tributary* is a river that joins a larger river. Find five tributaries, and print T on them.

7. *Downstream* is the direction of river flow from high places to its mouth. Draw an arrow pointing downstream on all the rivers on the east side of the Arapaho River.

8. *Upstream* is against the flow of a river. Upstream is toward the source. Draw an upstream arrow on or next to all the rivers on the west side of the Arapaho River.

9. Is the Iroquois River inside or outside the Arapaho Rivershed?

_____

10. What is the only river besides the Arapaho River that flows directly into the ocean?

_____

11. In this question use your imagination. What does the Arapaho Rivershed resemble that is also a part of nature?

_____

12. Color the ocean and the rivers blue.

MORE ABOUT WATERSHEDS I

Suggestions for Teaching

1. Orally read the information following "Suggestions for Teaching." During or after the reading, ask your students questions: e.g., "We all live in watersheds of one kind or another. What watershed do you live in?" Your students may answer by naming a large river such as the Delaware River. In such a case ask them to narrow their answer: e.g., the Rancocas River, which is a tributary to the Delaware River.

2. On the board write the heading, "Effects of Top Soil Loss." Then, as you mention the losses (listed below) write them on the board. Your students may copy the listing and take notes as you move through the lesson.

a. Crop reduction
b. River channel destruction
c. Water contamination
d. Lake sedimentation
e. Turbine damage
f. Wild/plant life destruction
g. River mouth clogging
h. Lake pollution

Watersheds: Everybody's Concern

Most water problems begin in somebody's small watershed. Many careless acts in many small watersheds grow into very large problems. Likewise, many careful efforts in many watersheds, both small and large, will prevent problems from arising.

In fact, starting small is the most effective way to manage our water resources. Our national government recognized this in one of its publications: "Water problems . . . are local problems. . . . Problems of water shortage, floods, pollution, or sedimentation must be met within the confines of each watershed."

How Water and Soil Problems Build Up

Here is an example of how individuals contribute to the making of great problems:

Farmer Jones is careless about the way he takes care of his land. After he harvests his crops he fails to plant a cover crop. Then the hard winds and rains of winter sweep over his land. Tons of soil are blown or washed into ditches, brooks, and rivers. With the soil goes minerals, such as nitrates, that are important to the growth of plants.

Multiply Farmer Jones' soil loss by the losses from thousands of other farms. Add to this the loss of soil from excavations for new homes and bulldozing for highways. Include the soil washed away from unplanted ditches, overgrazed pastures, and even small children's mud puddles. The total yearly loss in the United States? Enough topsoil to fill a convoy of dump trucks stretching to the moon—some 3 billion tons.

Effects of Top Soil Loss

What are some of the effects of topsoil washing off the land?

a. Topsoil loss from farms brings about a reduction in the size of crops. Plentiful topsoil with lots of nutrients produces large crops of fruits, vegetables, and grains.

b. Much of the washed-away soil settles on the bottoms of river channels. Thus, the depth of the water in channels is decreased. This means that large boats may not be able to navigate the river. It also means that during periods of heavy rain the rivers will not be able to hold as much water. The rivers may then flood and cause great damage to property. Often, in such situations, lives are lost.

c. Soil, or *sediment*, as it is often called, will increase the expense of filtering water to be used in homes or businesses.

d. Sediment settles in the bottoms of lakes and reservoirs, particularly near dam sites. This reduces the capacity of the reservoir to hold water. This, in turn, reduces the ability of the dam to generate electricity. Also, in times of heavy rain, the reservoir will not be able to hold as much water, and floods will occur from the spill-over. In times of little rain, there may not be enough water for homes, factories, and farms.

e. Sediment-filled water, falling from great heights, can damage the blades of the huge turbines at the bottoms of dams.

f. Sediment in lakes and streams covers fish nests, fish eggs, and fish food supplies. Sediment may eventually destroy recreational and commercial fishing. Sediment can cover and pollute clam and oyster beds.

g. Sediment that eventually reaches the sea is deposited at the mouths of the rivers. The sediment clogs bays, harbors, channels, and canals. It can affect the direction of flow and speed of currents. It can change the force and direction of waves.

h. Sediment that washes into lakes carries many different minerals such as calcium, salt, lime, and manganese. These minerals may pollute lakes to the extent that plants and fish are destroyed.

MORE ABOUT WATERSHEDS II

1. _____   2. _____

_____   _____

The two illustrations show a sequence that happens all too often to American farmers. Study the pictures, then fill in the blanks with words from the list at the bottom of the page.

Picture 1:

1.   The farmer has been farming this land for years, but he neglected to plant _____ crops after each growing season's _____. Year after year, topsoil _____ away and was carried to _____ and _____. Eventually, _____ were created by the _____ and _____ of winter. Each year the farm yielded _____ crops and _____ cash income. The family could not afford to buy new _____ or pay bills for such things as _____.

2.   Think of your own adjective to describe each of the nouns that follow. _____ gully, _____ rain, _____ farm

Picture 2:

1.   As the farmer became poorer, he was unable to keep up repairs on his _____, _____, and _____. No one wanted to buy his farm; there was no way the _____ could be replaced, except at great expense. During a family council it was decided they would _____ the farm and move to the city in the hope that jobs could be found. They piled their few belongings into the old pickup truck and moved out, never to return again.

2.   You will notice that there are two blank lines under each picture above. Compose your own caption—a sentence—that tells the main idea of each picture.

3.   Write what you think the two boys are thinking as they read the sign and look at the pond.

_____

_____

_____

_____

abandon	cover	fewer	house	rivers	topsoil
barn	electricity	gullies	less	silo	winds
clothing	eroded	harvest	rains	streams	

125

THE WATER CYCLE: AN EXPLANATION

Background

There is a saying, "What goes around, comes around." This bit of wisdom could have no better application than to the hydrocycle, or water cycle. The precipitation that falls to Earth—rain, snow, hail, sleet—eventually makes its way back to the atmosphere. Then, it falls once more. And, this has been going on for thousands of years.

Although the hydrocycle is constant, it is not always consistent in terms of where and when the precipitation occurs. Some areas of Earth may have sufficient precipitation to sustain crops, to sustain life for long periods of time. Then, suddenly, a drought might occur. Or, so much precipitation might fall that floods are caused. As a result, humans have learned to store water in such places as reservoirs for times of water scarcity. Humans have also built dams to hold back water in times of heavy precipitation.

Following are some facts relative to how the hydrocycle works:

1. Heat energy from the sun draws water, in the form of water vapor, from Earth to the atmosphere. This process is called *evaporation*. Moisture evaporates from a great number of sources: large bodies of water such as oceans and lakes; river and streams; animals, including humans; exhaust from machines such as automobiles, airplanes, and tractors; swamps; plants; forests; snow fields and glaciers.

2. As evaporated water reaches the atmosphere it cools and forms into very tiny droplets. The droplets join other droplets, and clouds are formed. When conditions in the atmosphere are right, the heavy droplets fall to Earth in various forms of precipitation.

3. The water that reaches Earth takes various "paths." Most falls into the oceans and other bodies of water. Other water falls on hard surfaces such as pavements, stony ground, etc., then "runs off" to rivers and streams, and eventually finds its way to the oceans. A great quantity of water sinks into the ground, where it may be stored for long periods of time. Much of the underground water, however, will drain to the oceans via underground routes.

4. It is fortunate that ground water does not float off or evaporate immediately because ground water is needed by plants, trees, and animals. The ground water beneath the surface may be reached by holes dug or bored into the ground (wells). It should also be noted that natural springs flow because water at a higher level is exerting pressure on the water below and causing it to rise. It is entirely possible that water reached by deep wells has been stored under ground for hundreds, if not thousands, of years. Finally, some ground water is evaporated by the heat of the sun, if the water is within a few inches of the surface.

5. All in all, we see a cycle: Precipitation falls, moisture evaporates into the atmosphere, the evaporated water cools and forms into clouds, clouds release droplets, the droplets fall to Earth as precipitaton, and the cycle begins all over again.

Suggestions for Teaching

Make a transparency of the diagram of the water cycle and project it. Explain the movement of the water vapor from the ocean (1) to the atmosphere. The wavy lines represent the rising vapor. The moisture that is transformed into vapor is coming from various sources, for example, the animals and tractor. Explain that the advancing air mass (2) is blowing the vapor and that the vapor will form into clouds (3, 4). The moisture collected in the clouds will become heavy and fall to Earth in various forms of precipitation (5). Finally, the precipitation will either run off (6), evaporate quickly (1), be stored underground (7), or move to the ocean (7).

Note: It is important to explain the water cycle using the diagram because the following activity presumes the students understand the cycle. An alternative to an oral explanation would be to photocopy this page, distribute it, and have students study the material. If this procedure is followed, this paragraph can be deleted from the photocopy.

UNDERSTANDING THE WATER CYCLE

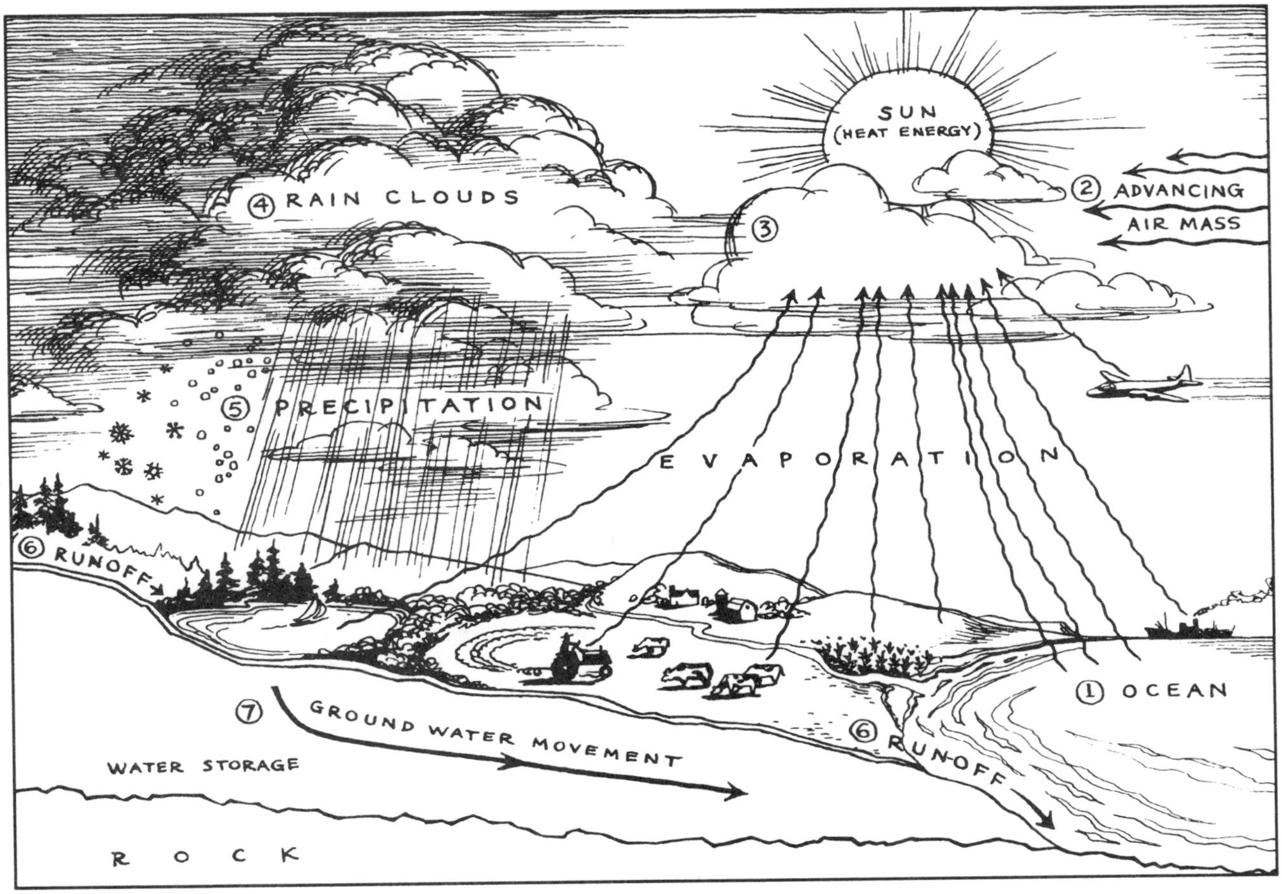

Water Cycle

The diagram shows how water is *recycled* from Earth to the atmosphere and back again to Earth. The recycling takes place again and again; it has been taking place since Earth was formed. The recycling is called the **water cycle**, or the **hydrocycle**, which is the more scientific name. *Hydro* means water.

To Do:

1. The wavy lines show moisture, in the form of vapor, drawn upward by the heat of the sun. Name six things from which moisture is being drawn. The airplane is one, and the ocean is another.

a. Ocean   e. _____

b. Airplane   f. _____

c. _____   g. _____

d. _____   h. _____

2. Print CLOUD FORMATION in number 3 of the diagram.

3. Notice how dark the clouds are in number 4 of the diagram. They are getting ready to drop **precipitation** (forms of water) on Earth.

What three forms of precipitation are shown?

_____

_____

_____

4. Number 6 in the diagram is labeled RUN-OFF. What is one of the things that the water is flowing

into? _____

5. Number 7 shows ground water flowing toward the ocean. And, some of the water is stored in the ground. Show the water that is stored by drawing small circles and dots (o.°o.°o) in the area.

6. Show the rock layer with diagonal (//////) lines.

WATER FROM UNDERGROUND SOURCES I

Background

Following are some "water vocabulary" words that students should understand:

☐ *Well*: This is created when a hole dug in the earth reaches a water-holding layer. A *dug well* is formed by digging a hole that is about four or five feet in diameter. A *bored well* is one that is dug by a powerful drill, much as a carpenter's drill and bit cuts a hole in wood. The well hole is about 6" in diameter. After the hole is bored, a pipe is fitted in it, and an electric pump is installed at the bottom of the hole.

☐ *Artesian Well*: In this type of water supply the water flows out of the earth naturally. It flows on the principle that water always seeks its own level. The flow is caused by water at a higher level pressing on the water below. Usually, the water begins to flow after some digging has created an opening. However, some artesian wells flow through a natural opening in the earth.

☐ *Spring*: This is also a flow of water from a natural opening. Springs are found in many different places: valleys, slopes, plains, mountains, and deserts. Many lakes receive their water from springs that are below them.

Additional Information About Ground Water

• Some spring water travels hundreds of miles before escaping to the surface. In desert regions, for example, springs may flow from ground on which rain rarely falls because the water has traveled underground from distant places.

• The temperatures of spring waters vary greatly. Water that comes from deep in the earth, where the rock is hot, will have high temperatures. People in Iceland, for example, heat their homes with water from nearby springs.

• Australia may be the world's greatest user of artesian wells. Thousands of sheep and cattle there drink water that flows naturally from holes in the ground. Without its artesian wells, one-third of Australia's interior would be practically useless for ranching and farming.

• Some of the water that drains into the earth sinks only a few inches. This water may soon evaporate, or it may be absorbed into the roots of plants and trees.

• Much of the water that strikes the earth continues to move downward until it cannot go any further because it has met impervious rock. Then, the water begins to accumulate and rise upward. The top of the underground reservoir that forms is called the *water table*. The water in the reservoir is called ground water. The water table may be significantly lowered during times of drought when wells are constantly giving up water that is not replenished. It is not unusual for water tables to lower as much as 50 feet and more. When this occurs, the well pipes cannot reach the water.

• When fresh water is drawn from the ground in coastal areas, salt water from the ocean may take its place. Of course, salt water that is drawn from wells is of little use unless the salt is removed.

• The underground reservoir is not a great hollow filled with water. Rather, it is a layer of sand, gravel, or porous rock that holds water as a sponge might hold it.

• A lowering of the water table may cause a sinking of the land above it. Roads might cave in, foundations might crack, and in-the-ground pipes might break. When enduring a prolonged drought, people who obtain their water from wells must use as little water as possible.

Suggestions for Teaching

1. It would be helpful to convey the information on this page to students before they complete the activity on the next page. Encourage them to take notes. Note-taking increases retention because students are not only listening, but also actively responding.

2. Distribute copies of the underground water diagram. Explain that it is a cross-section diagram. Call attention to some of the details of the diagram as a means of reviewing the information on this page.

Name: _____   Date: _____

WATER FROM UNDERGROUND SOURCES II

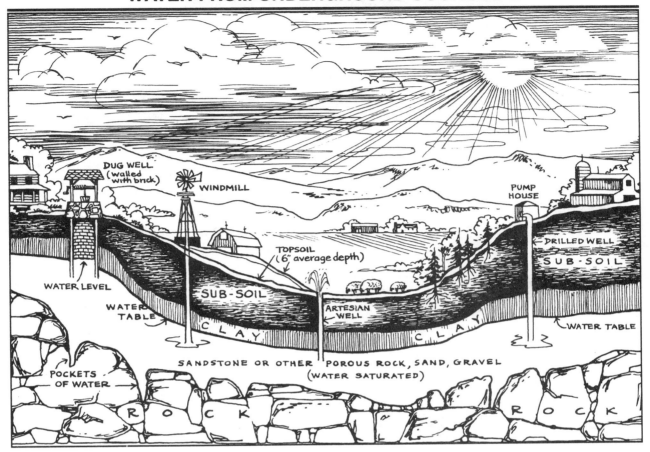

Many people in the world, especially city dwellers, get their water from reservoirs, which are very much like lakes. Some reservoirs are natural, and some are made by humans. Also, many people get the water they need from rivers that flow close to their cities and towns. However, millions of people obtain their water from underground sources such as springs and wells. And, that is what the diagram above shows.

To Do:

1. The diagram shows four ways to get water from underground sources. Starting from the left of the diagram, find them and list them below.

a. _____   c. _____

b. _____   d. _____

2. Study the picture of the dug well. How is the water brought to the surface?

_____

3. Why might the windmill fail to bring water to the surface? _____

_____

4. According to the diagram, what prevents the water in the *water-saturated* layer from sinking further into the ground?

_____

_____

_____

5. What event in nature might cause the water table to lower and not provide water for the wells?

_____

6. What layer prevents the water in the water table from rising higher, toward the surface?

_____

7. The water table is not a great hollow in the ground that is filled with water. The space is filled with sand, gravel, and porous rock. It holds water in much the same way a sponge holds water.

Draw droplets of water (o₀ₒ°o) throughout the water-saturated area.

Name: _____ Date: _____

SOURCES OF FRESH WATER: SURFACE I

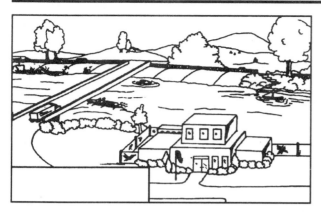

1. Some cities get water from rivers. Trenton, New Jersey, draws water from the Delaware River.

2. Other cities get water from nearby lakes. Chicago, Illinois, uses Lake Michigan water.

3. Water from lakes or rivers is first drawn through *water intakes*. A screen over the intake blocks fish, sticks, etc. The water is then piped to a water treatment plant.

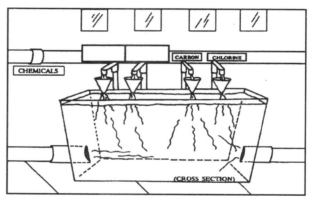

4. In the treatment plant, chemicals are added to the water to destroy harmful bacteria and to eliminate disagreeable odors.

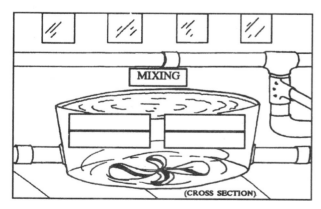

5. The chemicals are thoroughly mixed in great mixing basins.

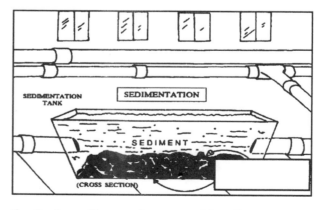

6. In the *sedimentation room*, bits of matter settle to the bottom of the tank. The collected material is called *sludge*.

SOURCES OF FRESH WATER: SURFACE II

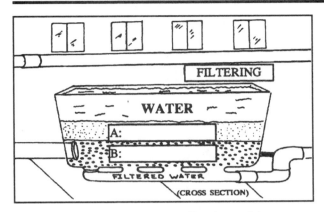

7.   In the filtering tank, very fine impurities are removed as the water filters through layers of sand and gravel.

8.   As a final precaution, chlorine gas is added to the water to kill any remaining bacteria.

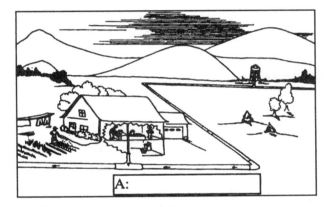

9.   The purified water may be pumped into high tanks for storage. When a valve in the tank is opened the water flows into the pipes at a lower level.

10. Water under pressure fills large water mains and the smaller pipes of houses and other buildings.

1.   Complete the drawings as follows:

#1: Print *WATERWORKS*. Color the river blue.
#2: Color the lake blue and the large tree green.
#3: At A, label the screen as *WATER INTAKE*.
#4: Label the first can *ALUM*, the second *LIME*.
#5: Print *ALUM, LIME, CARBON, CHLORINE* in the mixing tank.
#6: Print *SLUDGE* by the arrow.
#7: At A print *SAND*; at B print *GRAVEL*.
#8: Print *FINAL CHLORINATION* in the sign on the back wall.
#9: Print *STORAGE TANK* in the blank.
#10: Print *WATER MAIN* at A.

2.   From what two sources might cities draw water?

_____   _____

3.   What prevents large objects from entering the water intake? _____

4.   What are two reasons why chemicals are added to the water in the purifying process?

_____

_____

5.   What is the final step in removing fine particles from the water? _____

_____

6.   What is the purpose of the final dosage of chlorine? _____

_____

7.   Why is the water stored in a tank higher than any of the houses and other buildings?

_____

_____

WATER POLLUTION: SPILLED OIL I

Petroleum—oil—is a major polluter of water. Sometimes oil leaks from tank ships when they collide with other ships, or when they run aground. Large amounts of oil are released into waters when tankers take on or discharge oil in ports. Sometimes tankers illegally clean out their tanks in ocean waters. In a recent year 185 tank ships, 338 tank barges, 5220 other vessels, and 3929 non-vessels were involved in oil-leakage incidents in and around United States waters.

Oil also leaks into waters from shore installations—a pipe in a refinery might burst, or an oil transportation truck might have an accident while loading oil at a dock. There have been cases when railroad tank cars have overturned and spilled oil into rivers, which eventually carried the oil to the sea.

Oil spills kill fish, water birds, and sea animals such as sea otters and seals. When sea birds and sea animals become covered with oil their chances of staying alive are not great.

Activity

Use the figures at the bottom of the page to have your students make one bar graph showing the number of oil spill incidents within the five-year period from 1989 to 1993 and a second bar graph showing how many gallons of oil were spilled in those incidents.

1. On a transparency or on the board write the figures on oil spill incidents and gallons of oil spilled.

2. Explain how the figures have been rounded—to the nearest 500 in the spill incidents and to the nearest 500,000 in the number of gallons spilled.

3. Explain the footnotes, which tell the source of the information and other pertinent facts. Students should be told that, even though the information is the latest available, statistic-gathering organizations are sometimes as much as three years behind because it take time to gather statistics from all the reporting organizations.

4. Have your students show the incidents graph in another way, that is, in a picture graph. Each 1000 incidents could be shown by a tanker ( ). Since they will have already worked on the bar graphs on the page, they should be able to make a title of their own.

5. Some questions to ask: (*Note*: Use rounded figures.)
a. How many more gallons of oil were spilled in 1989 than in 1990–1993 combined? (6,500,000 gallons)
b. What might account for the fact that so much more oil was spilled in 1989 than in any of the following four years? One very large spill, for example, the Exxon Valdez spill in Alaska in 1989, spilled 10,080,000 gallons.

Oil-Polluting Incidents Reported in and Around United States Waters 1989–1993 [1]				
Year	Incidents	Incidents Rounded [2]	Spilled Gallons	Spilled Gallons Rounded [3]
1989	8,562	8,500	25,531,292	25,500,000
1990	10,186	10,000	13,907,783	14,000,000
1991	10,405	10,500	2,156,063	2,000,000
1992	9,131	9,000	1,572,341	1,500,000
1993	9,672	9,500	1,543,578	1,500,000

Notes: [1] Data from *Statistical Abstract of the United States*, 1998, p. 241 (latest information available)
[2] Incidents rounded to nearest 500
[3] Gallons rounded to nearest 500,000

WATER POLLUTION: SPILLED OIL II

Oil-Polluting Incidents Reported In and Around United States Waters
1989–1993

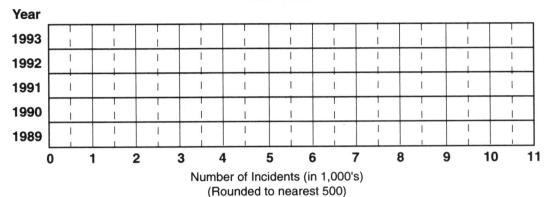

Number of Incidents (in 1,000's)
(Rounded to nearest 500)

Amount of Oil Reported Spilled In and Around United States Waters
1989–1993

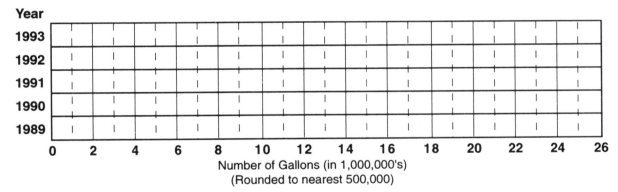

Number of Gallons (in 1,000,000's)
(Rounded to nearest 500,000)

Title: _____

_____

Year									
1993									
1992									
1991									
1990									
1989									

Key: Each symbol (⬛) represent 1,000 oil-spill incidents

OIL SPILL IN THE SHETLANDS: AN ENVIRONMENTAL DISASTER I

On December 3, 1992, the world was shocked to read about a 23-million-gallon oil spill off the northern coast of Spain near Coruna. The oil spilled into the Bay of Biscay, coating 50 miles of beaches and hundreds of square miles of the bay with sticky, gummy, foul-smelling oil. Thousands of oil-soaked birds and fish were killed or injured.

To Do:

1. On the map, label the *Atlantic Ocean* at A, *Europe* at B, the *Bay of Biscay* at C, *Spain* at D, *Portugal* at E.

2. What are the facts of the Coruna oil spill?

a. Date of Coruna disaster? _____

b. Place where oil spilled? _____

c. Amount of oil spilled? _____

d. Effect on coastline? _____

_____

e. Effect on birds and fish? _____

_____

One month after the Coruna disaster, an even more terrible sea tragedy occurred north of Great Britain. A tanker carrying 25 million gallons of crude oil became lodged on the rocks off the southern coast of the Shetland Islands. Winds of hurricane force had blown the ship there after its engines failed. The *Braer*, as the ship was called, was transporting the oil from Norway to Canada.

3. Label *England* at F; *Scotland* at G; the *Shetland Islands* at H; *North Sea* at I; *Norway* at J.

4. What are the facts of the Shetland oil spill?

a. Location of the Shetland Islands?

_____

_____

b. Cause of accident? _____

c. Amount of oil spilled? _____

Following are some additional facts and results of the Shetland Islands spill:

• All 34 of the crew were rescued from the sinking ship as it was being pounded against the rocks.

• The *Braer* was a single-hulled tanker. If it had been double-hulled, there is a strong possibility that the oil might have been contained on the ship. *Note*: The *hull* is the sides of the ship.

5. Thought question: Why would a double-hulled tanker have been less likely to spill oil?

_____

_____

• Some of the most surprising aspects of the spill were the effects it had inland. Usually, it is just the beaches and the sea water that are damaged, but the 80-m.p.h. winds blowing toward the land changed that. One islander was quoted in a news report as saying, "It's (the islands) always been a very healthy place. I think what surprised us all is the effect the oil spill has been on agriculture and livestock." The fierce winds carried oil spray inland, smearing the land, the homes, the animals, and the people. The oil spill had many unfortunate results. Drinking water was contaminated. Sheep and other grazing animals were moved from coastal areas because the grass was oil-covered. The animals had to be fed hay and other dry food. The famous Shetland ponies were threatened. The salmon fisheries were in danger. This was doubly discouraging because Shetland salmon are especially desired because the waters from which they come are known for their cleanliness and purity. Hundreds of birds and fish were killed. One photograph made at the scene showed an oil-soaked bird plucking an oil-drenched fish from the sea. A colony of seals was in serious danger from the oil that covered them and the food they needed to survive.

6. Underline the following:

a. The speed of the winds

b. What happened to drinking water

c. Why sheep and other animals were moved from coastal areas

d. Why Shetland salmon fisheries are so well known

e. Why seals were in serious danger

OIL SPILL IN THE SHETLANDS: AN ENVIRONMENTAL DISASTER II

© 2000 by The Center for Applied Research in Education

The site of the oil spill (60°N) is only 450 miles from the Arctic Circle (66½N). But, the islands are warmed by the North Atlantic Drift, which is a current of warm water. There was the danger that the Drift would carry the oil northeast toward Norway. Places and wildlife hundreds of miles from the original spill would have been in danger of contamination from the oil.

To Do:

7. Label the *North Atlantic Drift* at **K** on the map; *60°N* on the line of latitude labeled **L**; *Arctic Circle* on the line labeled **M**.

8. Circle the sentence that tells why the oil spill was a possible danger to the coasts of Norway.

9. Following is an opportunity to put your latitude and longitude skills to good use:

a. What is the approximate latitude and longitude of the Shetland oil spill?

Latitude: _____ Longitude: _____

b. Print *Iceland* on the island just south of the 66½° line of latitude and west of the 10° line of longitude.

c. Print *France* in the country north of Spain that is mostly south of the 50°N latitude line and mostly east of the 0° longitude line.

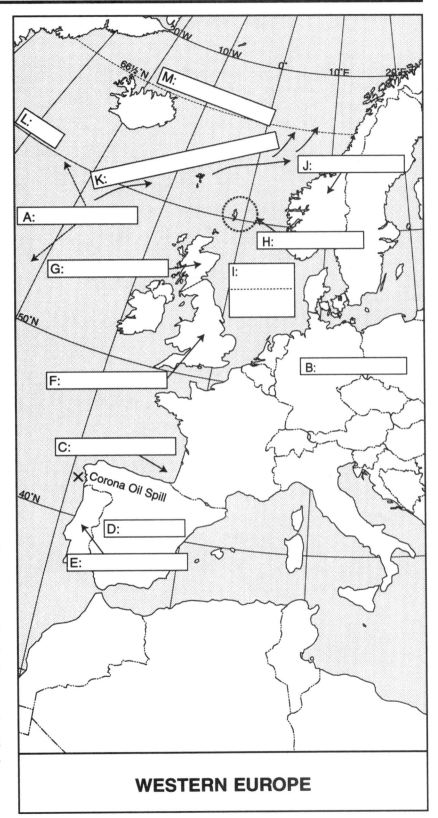

WESTERN EUROPE

THERMAL POLLUTION: WHAT IS IT? HOW CAN IT BE CONTROLLED?

Water that is used to cool the giant machines of power plants and industry becomes very warm. When this warm water returns to its source, it mixes with the cooler water there. The effect of this mixing is to increase the temperature of the lake, stream, or river that receives the water.

The addition of heat to natural waters to the extent that plant and animal life in the water are harmed is called *thermal pollution*. Following are some of the undesirable results of thermal pollution.

• *Adult fish can be destroyed*. As heat is added to water two things occur: one, the oxygen content of the water is reduced; two, with an increase in heat, fish need much more oxygen to survive. If the oxygen content of the water becomes too low, the fish will die of suffocation.

• *Fish eggs and baby fish may be destroyed*. Water temperature helps determine when fish eggs will hatch. If warm water is discharged into streams too early in the year, some fish may hatch too soon. The food that the newborn fish must have may not be available. The baby fish will then die of starvation.

• *Undesirable plant life in the water is increased*. Surface plants, **algae**, grow more rapidly with heat. The water surface soon becomes covered with slimy algae, and foul odors begin to rise from the water. The algae consumes even more oxygen, which means even less oxygen for the fish.

To Do:

1. Underline the sentence that

a. defines thermal pollution.

b. tells what happens to fish when oxygen becomes low in water.

c. tells how fish that are born too early can die of starvation.

_____

_____

_____

_____

_____

d. tells how surface plants, *algae*, are helped to grow.

2. The drawings at the top of the column show two ways hot water can be cooled before being discharged into streams. Write the first sentence on the lines below picture 1. Then, write the second sentence on the lines under picture 2.

1st: As the warm water leaves the factory, it can be sprayed into streams. The air will cool the water.

2nd: Send factory-hot water through many outlets. The water will cool and will not be concentrated in one part of the stream.

3. Think of one other way to cool water discharged by factories.

_____

_____

_____

AIR: WHAT IS IT?
HOW IS IT POLLUTED?

11-1   All About Air I (Instructor) 138

11-2   All About Air II 139

11-3   Demonstration: Proving That Air Has Weight (Instructor) 140

11-4   Understanding the Atmosphere 141

11-5   What Is Air Pollution? I (Instructor) 142

11-6   What Is Air Pollution? II 143

11-7   The Effect of Acid Rain on Green Plants (Instructor) 144

11-8   Acid Lakes: What Can Be Done About Them? 145

11-9   Air Pollution: Industry and Power Plants I 146

11-10 Air Pollution: Industry and Power Plants II 147

11-11 Polluted Air Is Expensive I 148

11-12 Polluted Air Is Expensive II 149

11-13 A Polluted Air Disaster 150

ALL ABOUT AIR I

Air Is a Mixture of Gases

Nitrogen gas makes up about 78% of air. This gas is vital to the growth of plants and animals. Nitrogen enters the soil through the action of rain. Plants absorb the nitrogen from the soil. Most animals get the nitrogen they need by eating plants. Humans and other meat-eating animals get the nitrogen they need by eating both plants and animals.

About 21% of air is oxygen. No plant or animal could live without oxygen. When air is drawn into the lungs, the nitrogen is not used and is exhaled; but the oxygen is carried by the blood to all parts of the body.

Another gas known as argon makes up about 1% of air, and other gases are present in air in smaller amounts.

Air Contains Water

The moisture in air is called **water vapor**. The water particles in water vapor are so small that they are not visible. However, as air gets colder, the water particles will unite with each other. When they reach the size of droplets, they will fall to the earth as rain, snow, sleet, or hail.

Air Contains Solid Particles

All "normal" air contains solid particles such as pollen from plants, microbes, dust, and salt from the sea. In fact, a cubic inch of air may contain as many as 100,000 particles. But the solid particles in a cubic inch of "dirty" air may number in the thousands. For example, smoke is easily seen because it contains millions more particles than clean air.

Air Has Weight

Imagine a column of air with a base of one square inch stretching far into space. At sea level, that column would weigh about 14.7 pounds. The column of air has weight because it is made up of gases, water vapor, and solid particles. At higher levels, air weighs much less. At 20,000' it weighs about 7 pounds per square inch.

Air Expands When It Is Warmed and Contracts When It Is Cooled

Air is made up of billions upon billions of **molecules**. A molecule is a very small bit of matter—so small that a grain of sand would be thousands of times larger. These molecules are in constant motion. When they are heated, they become even more active. They begin to spread out, expand, and therefore need more room.

When air molecules are cooled, they become less active. They begin to draw together, contract, and need less room to move about.

Because air expands when heated and contracts when cooled, a football will shrink when left out on a freezing night. The molecules of cold air in the ball use less space. If the football is left in the warm sun, it will return to its original size.

In the balcony of a theater it may become uncomfortable because it is too hot. On the ground floor it probably would be cooler. This is explained by the fact that warm air rises while cold air settles.

This same movement of air takes place in the earth's atmosphere. As the sun heats the earth, it also heats the air near it. The heated surface air expands and rises. Then cooler air moves in to take the place of the departing warm air.

As the warm air reaches higher altitudes, it begins to cool. So, down it settles to take the place of air near the surface that is being heated. Thus, air is in continuous circulation vertically, that is, up and down.

Suggestions for Teaching

The questions on the facing page are based on the information on this page. You may want to use the questions as a study guide—that is, as you present the information the students respond to the questions. Or, you may choose to photocopy this page and the facing page. If so, your students can read the information and respond to the questions.

ALL ABOUT AIR II

1.   Before each statement below write a T for True or F for False.

a. _____ Nitrogen enters soil from the air.

b. _____ Plants absorb nitrogen from the soil.

c. _____ Meat-eating animals, including humans, absorb nitrogen from the air.

d. _____ Rainfall helps nitrogen enter soil.

e. _____ The nitrogen we inhale into our lungs is used by our bodies.

f. _____ Plants are able to absorb nitrogen directly from the air.

g. _____ Both plants and animals must have nitrogen in order to live.

h. _____ Water vapor forms into water droplets when air becomes warm.

2.   Air contains a number of solid particles, including:

_____

_____

_____

_____

3.   The higher one goes into the atmosphere, the (less / more) air weighs.

4.   Molecules in air begin to move much more rapidly when heated. To show this increased action, draw an arrow on each of the "molecules" (•••) shown in the diagram of the hot air balloon. Since molecules move in all directions and bombard the sides of the containers that hold them, show each arrow as pointing in a different direction.

Several of the molecules have arrows on them to show you how to get started. (You should realize, of course, that there would be millions more molecules in the balloon than are shown.)

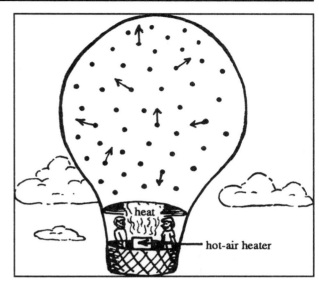

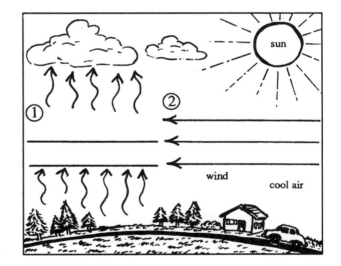

5.   When you complete the diagram below, it will show that warm air rises when heated and is replaced by cooler air. This action is one of the causes of wind.

Print the following in the diagram:

1.   *Warm air rising*

2.   *Cool air replacing the rising warm air*

DEMONSTRATION: PROVING THAT AIR HAS WEIGHT

Background

Visualize a column of air pressing on one square inch of sand at sea level. The column stretches far into the atmosphere. What is the weight of the air in the column? Almost 15 pounds. The weight is there because there are gases, dust, pollen, and other small particles in air. Also, air at higher altitudes presses on the air below it and adds weight. We don't feel the weight because air presses on us equally from all sides.

1. Objective

The students will understand that air has weight and that air has the capacity to crush objects under certain conditions.

2. Introduction

Pose the question: Suppose that for some reason you wanted to crush a can. What are some ways it could be done? (Accept all suggestions.) Do you think it might be possible to crush it with air?

Obtain a gallon can with a screw-on cap, a hot plate, a tablespoon of water, and paper towels. *Caution*: Be sure the interior of the can is completely clean of any liquid it might have held, and be sure all paper (labels, etc.) is removed from the outside of the can.

Remove the cap from the can and pour the water into the can.

Place the can on the hot plate.

After all the air (in the form of vapor) has been driven from the can, quickly replace the cap on the can. *Caution*: Use the paper towels as insulation when touching or picking up the can.

Remove the can from the heat and allow it to cool.

Observe: The can will crush.

Explanation: The air pressure on the outside of the can was greater than the air pressure within the can; therefore, the can crushed.

Extension question: What was the total weight of the air on the can? (Find the area in square inches of all six sides of the can, then multiply by 15 (the approximate weight of air at sea level.)

Show Figure 1 via a transparency projection. By doing so the demonstration is shown in another way. Perhaps students can use the transparency to make their own drawings.

Note: Figure 2 will help students to understand that air weighs less at higher altitudes.

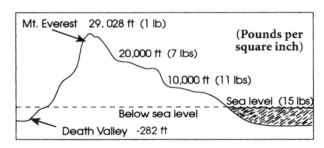

Figure 2

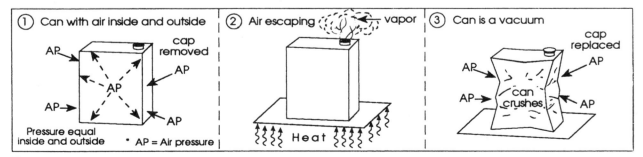

Figure 1

UNDERSTANDING THE ATMOSPHERE

The Earth's Atmosphere and Its Layers of Air

Air pollutants have a tendency to settle. Since most human activities occur within the first 50 feet above the surface of Earth, it is important that this portion of the atmosphere be kept clear of harmful particles and gases.

EXOSPHERE
from 400 miles to outer space
• There is practically no air in this layer.

IONOSPHERE
from 50-400 miles up
• The air in this layer is very thin, especially at the upper levels. The temperatures at the upper levels reach 2000°F.

MESOSPHERE
from 20-50 miles up
• The Mesosphere is sometimes considered to be a part of the Stratosphere. At lower levels temperature is about 30°F; upper levels, −100°F.

STRATOSPHERE
20 miles (approximately)
• Strong, steady winds at lower levels. Very little wind at upper levels.
• Contains a layer of special oxygen (ozone) starting about 15 miles up. This layer protects the earth from harmful rays of the sun.

TROPOSPHERE
10 miles (approximately)
MT. EVEREST: Earth's highest point 5½ miles high
SEA LEVEL
• Almost all weather occurs in this layer. Winds blow in all directions, up and down, east-west-north-south.
• Temperature falls about 3½° for every 1000' rise in elevation.
• About 90% of the weight of the atmosphere is in this layer.

Earth

To Do:

1. As you can see in the diagram above, Earth has layers of air surrounding it.

How many layers are shown? _____

2. What are the names of the layers starting from the layer closest to Earth and going upward?

Closest layer? _____

2nd closest: _____

3rd closest: _____

4th closest: _____

5th closest: _____

3. In which layer is Earth's highest mountain, Mt.

Everest? _____

4. In which layer does the most weather occur?

_____

5. Which layer has special oxygen (Ozone) that protects Earth from the sun's harmful rays?

_____

6.a. How many degrees does the temperature drop every 1000' into the Troposphere?

_____ degrees

b. Imagine that at Earth's surface the temperature is 80°F. What would the temperature be at an altitude of 5000'?

_____ degrees

7. How many miles into the atmosphere would a rocket have to go before it reached the Exosphere?

_____ miles

8. Why is the Troposphere more polluted than any of the other layers?

_____

_____

WHAT IS AIR POLLUTION? I

"Normal" Air and Polluted Air

Normal air contains several gases, water vapor, and certain solid particles. Generally, none of these things are harmful to humans, animals, or plants when they are held in proper balance. In fact, all of the ingredients of air, including the solid particles, are necessary if nature is to do its work.

Here is an example of the need for solid particles in air: Water vapor molecules condense (change to liquid form) when they rise into the atmosphere and cool. But, when the molecules begin to condense, they condense on something. That "something" is one of the tiny solid particles in the air.

However, if the air is made to contain solid particles and strange gases in such quantities that they are harmful to humans, animals, plants, or property, the air is said to be polluted.

How Does Nature Pollute Air?

Nature itself sometimes pollutes air. When a forest fire starts from natural causes, such as lightning, quantities of soot particles and gases are shot into the air. Volcanic eruptions send ash and gases into the atmosphere. Dust storms in desert areas often load the skies with soil particles.

But, all such occurrences are part of nature's way of balancing the forces of the earth. As you have read, soil particles are important for rainmaking. Ash from volcanoes fertilizes soil. Even the ashes and charcoal that remain after a forest fire will be used by new green plants as a part of their food-making process.

There is a constant exchange among all the forces and elements of the earth. In this way all things in the earth, on the earth, and over the earth are kept in balance. Sometimes, however, the balance in nature is upset by humans through their activities.

How Do Humans Pollute the Air?

Humans are much more dangerous polluters of the atmosphere than nature. This is because human pollution is often concentrated in certain areas, such as cities and towns, for long periods of time. Helpful winds that blow away pollution or cleansing rain that washes out the skies may not occur often enough, or long enough. Meantime, the pollutants are at work causing deaths, illness, discomfort, and property damage.

The undesirable things dumped into the air may be placed in two groups. One group contains solids. The other group contains gases. In most cases of pollution, both types of pollutants are put into the air together. A smoke stack, an automobile engine, and a jet airplane flying overhead are all emitting (sending out) gases and solids at the same time.

Suggestions for Teaching

1. If you orally present the information on this page, make photocopies of the facing page for your students. As you move along in your presentation, your students can write answers to the questions. This approach is really a form of note-taking.

2. You may want to photocopy both pages and have your students read silently, answering the questions as they are sequenced. *Note*: It is probably more effective as a study technique for students to respond to questions as they read rather than after reading an entire selection. This is because each question provides a purpose for reading, as suggested in F. Robinson's SQ3R (Survey, Question, Read, Recite, Review) study approach. Also, students are actively responding to what they are reading.

Name: _____   Date: _____

WHAT IS AIR POLLUTION? II

1. What are some of the ingredients found in "normal" air? _____

_____

2. Why are particulates (tiny bits of dust, etc.) necessary for the formation of water droplets? _____

_____

3. How can we define "air pollution"? _____

_____

_____

4. What are three ways that nature pollutes air? _____ ,

_____ , and _____

5. What is one good result of a volcanic eruption? _____

_____ _____

6. What is a positive result of forest fires? _____

_____

7. How do humans upset the balance of "good" and "bad" that comes from nature's pollution? _____

_____

_____

8. What are the two major kinds of pollutants humans send into the air?

_____ and _____

9. The illustrations below show common ways that air pollutants are introduced into the air by burning. On the blank lines in each picture write a short title that tells how the air is being polluted by burning.

Title: _____

Title: _____

Title: _____

Title: _____

THE EFFECT OF ACID RAIN ON GREEN PLANTS

Title: Acid Rain: Destroyer of Plant, Soil, and Animal Life

Background: Rain water from the clouds is relatively pure. However, as the rain falls through the atmosphere it picks up pollutants emitted from automobiles, factories, and other sources. Some of the pollution is in the form of sulfuric acid and/or nitric acid. Rain water that is contaminated by these acids is called ***acid rain***.

Acids have the power to corrode metal, rock, wood—almost anything they contact. Drop some lemon juice—an acid—on a wooden table, and in a short time corrosion takes place. The facades of buildings, statues, bridges, and other objects are easily attacked by acid ran and will eventually deteriorate. The soft, tender leaves of trees have virtually no protection against acid rain.

Acid rain also contaminates the water it falls on, as in lakes and rivers. Given sufficient acid rain, fish in a lake will die because the acids will destroy their food supply. Also, the ingestion of acid waters can damage fish internally, especially their gills.

Experiment:

Purpose: To determine the effect of acid rain on green plants

Materials: ❑ Two clear plastic or glass jars with wide mouths and screw-top covers

❑ Litmus paper

❑ Litmus paper color-matching chart for determining acidity (obtainable in tropical fish stores, swimming pool outlets, hobby shops)

❑ Two green plants

Procedures: 1. In jar "A" collect rain water as it falls from the sky. Screw on the cover.

2. In jar "B" obtain water from a water faucet. Screw on the cover. (Jar "B" is a control.)

3. Dip litmus paper into jar "A." Match against the color chart. Results: Acid, alkaline, or neutral (pH balanced)?

4. Repeat step 3 for the faucet water.

Drawings:

Conclusions: 1. The rain water we collected proved to be (acid, alkaline, neutral).

2. The water we collected from the faucet proved to be (acid, alkaline, neutral).

Application: If the rain water is found to be acidic, place a few drops each day for five days on live plant leaves. Do the same with the pH balanced alkaline faucet water on the leaves of the other plant. Observe and compare the two plants.

Question: What might be the result of frequent acid rain falling on a forest? (Answer: Eventually, the forest could be destroyed.)

© 2000 by The Center for Applied Research in Education

ACID LAKES: WHAT CAN BE DONE ABOUT THEM?

Lakes are like huge bowls. When it rains they collect large amounts of water—and that water may be acid rain. A one-acre lake (43,560 sq. ft.) catches about 2263 gallons of water from one inch of rainfall. The amount of acid in the rain depends upon location. Lakes near cities probably receive more acid per rainfall than lakes far away from cities. This is because cities are likely to have large numbers of automobiles and factories—the big contributors to polluted air.

1.   How many gallons of water would be added to a ten-acre lake that received 2" of rain over a period of a week? Hint: 10 acres × 2263 gallons × 2" = _____ gallons

2.   Why are lakes near cities more likely to be acidic than lakes in remote places?

_____

_____

_____

How does the rain that falls into lakes become acidic? As the rain passes through the atmosphere it picks up sulfuric acid and nitric acid. These acids added to a lake can seriously affect the lake's plant and animal life. As lakes become more acidic, the first organisms to die are the smaller ones. This reduces the amount of food available for fish and frogs; then they die from lack of food. It is only a matter of time before all the animal life in a lake is gone.

3.   How does a lake become acidic?

_____

_____

_____

4.   Can you think of one other way that lakes become acidic other than from falling rain?

_____

_____

_____

_____

Lakes that have become acidic can be restored. Here is one way it is done: Helicopters carrying storage tanks of limestone fly over lakes and release the limestone (calcium carbonate) in the form of a spray. The limestone neutralizes, or eliminates, the acid in the water.

5.   Circle the two words that tell what happens to the acid in water when it is treated with limestone.

6.   In what other way besides using aircraft could the water be sprayed with limestone?

_____

_____

_____

_____

_____

The effect of limestone on an acid lake is remarkable. In a matter of hours the water is restored and can be used for swimming and other forms of recreation. Fish recover from the effects of the acid, and the eggs they lay are no longer contaminated by poison acid. Plant life begins to grow, and the normal food cycle resumes.

7.   Circle the sentence that tells how fish are helped by having limestone sprayed over an acid lake.

An organization called "Living Lakes, Inc." has been important in the effort to restore lakes. They have been successful numerous times. They like to use the "liming method," as it is called, because limestone dissolves easily, is inexpensive, is easy to handle, and is nontoxic (not poisonous).

8.   List four reasons why Living Lakes, Inc., uses the liming method to restore acid lakes.

a. _____

b. _____

c. _____

d. _____

AIR POLLUTION: INDUSTRY AND POWER PLANTS I

_____

_____

1 There are more than 310,000 manufacturing plants in the United States. Some of them are huge, covering acres of ground and employing thousands of people.

2 The factories are important to all of us. They produce the food, clothing, and other things necessary to life as we know it.

3 In order for manufacturing plants to operate, they must have power to run machines, move materials, and supply heat and light.

4 Industry is not the only user of power. We also use power in our homes and communities.

_____

_____

5 Power is produced in generating stations. Two ways that these stations provide power is through the use of falling water (hydroelectric power) and nuclear energy. In neither of these two ways is smoke a result of the power-making process.

6 A third way, and the most widely used way that power is produced, is by burning coal or oil. The heat produced by the burning converts water to steam. The steam has the strength to move wheels and gears that generate electricity.

7 In carrying out manufacturing processes and in making electric power through the burning of fuel, smoke is produced. The picture on the next page shows smoke coming from only one of the manufacturing plants in the country. What you cannot see coming from the chimneys are poisonous gases, fumes, and the particulates that are mixed with the smoke.

8 Each year factories and power plants throw some 43 million tons of pollutants into the atmosphere.

9 The problem is this: How can we keep our manufacturing plants operating and, at the same time, eliminate, or at least greatly reduce, the air pollutants they give off?

Ways to Control Pollution from Manufacturing and Power Plants

10 *Use higher grade fuel*—When coal and oil are burned, sulfur dioxide is formed. This gas is undesirable. It is colorless, but it has a bad odor, not unlike rotten eggs. This gas is harmful to humans, animals, and plants. It will even attack and wear away bricks, mortar, and metals.

11 How much sulfur dioxide results from burning is dependent upon the quality of the fuel. The lower the quality, the higher the sulfur content. Several states and communities now strictly regulate the grade of fuel used by industry. Reduced sulfur dioxide emissions from smokestacks has been the welcome result.

12 *Burn off harmful gases*—As undesirable gases emerge from smokestacks, they can be set on fire and, therefore, eliminated. To insure that smoke from the fires will not contribute to pollution, jets of air can be blown into the flames. The extra oxygen greatly increases the temperature of the fire and makes the flames smokeless.

13 *Collect particles before they enter the atmosphere*—One of the by-products of coal that has been burned is *fly-ash*. Fly-ash particulates may be as small as dust specks or as large as grape seeds. Do not confuse fly-ash with soot. Soot is composed of carbon. It gives smoke its blackness and can be burned. Fly-ash is highly resistant to burning and is therefore difficult to destroy.

© 2000 by The Center for Applied Research in Education

AIR POLLUTION: INDUSTRY AND POWER PLANTS II

14 There are several ways to prevent ash from blowing out of smokestacks. One way is through the use of electrostatic precipitators. These devices cause fly-ash to stick to plates as the particles are being blown toward chimney openings.

15 Another way to catch fly-ash and soot is through the use of filters. As the gases make their way through small filter openings, the particles are caught and later removed from the filters.

16 Every year sees improvements in smoke-stack control. Aside from helping to keep the atmosphere clean, these improvements have other benefits. Particles collected by such methods as those described above can be put to good use. Fly-ash can be used as an ingredient in cinder blocks and concrete. Even harmful gases can be made into useful products.

To Do:

1.   Notice that each paragraph in the story is numbered. Which of the topic headings at the end of this question best tells about paragraphs 1–4? _____ Which paragraph heading best tells about paragraphs 5–8? _____

A: *Producing Power*

B: *Ways to Save Power*

C: *Manufacturing and the Need for Power*

2.   Circle the numbers, words, phrases, or sentences in the story that tell:
•   the number of manufacturing plants in the United States.
•   why factories are important.
•   the two ways of producing power that do not also produce smoke.
•   the two most common fuels used to produce power.
•   the number of tons of pollutants sent into the air each year by factories and power plants.

3.   Underline the sentence in paragraph 11 that tells what some states and communities are doing to reduce sulfur dioxide emissions.

4.   Underline the sentence in paragraph 12 that tells why extra oxygen blown into a fire reduces air pollution.

5.   Why is fly-ash more difficult to destroy than soot?

_____

_____

6.   Underline the sentence in paragraph 14 that tells how an electrostatic precipitator works.

7.   How can captured fly-ash be put to good use?

_____

_____

8.   Make up an appropriate caption for the picture, and write it on the blank lines below the picture.

Challenge

What sources of energy besides those already mentioned could be used to produce power but would not produce smoke. Try to think of at least two.

_____

_____

_____

_____

_____

_____

_____

POLLUTED AIR IS EXPENSIVE I

Polluted air . . .

- breaks down rubber tires
- kills trees
- erodes stone and metal statues
- kills plants and flowers
- rots and cracks leather products
- peels paint
- damages glass
- pits aluminum
- rusts train rails and bridges
- defaces and dirties buildings
- kills cattle

- ruins car paint
- turns paper brittle
- eats away wire insulation
- damages playground equipment
- soils clothing, rugs, and drapes
- spoils fruit
- fades clothing and curtains
- tarnishes silverware
- damages machinery
- causes accidents
- cuts down on sunlight

The Yearly Bill

Some costs of pollution cannot be accurately measured. It is impossible to determine the decrease in value of a great statue through erosion caused by polluted air. How does the presence of heavily polluted air affect the value of land when one goes to sell it? What is the price of sneezing, coughing, or eye irritation due to dirty air? No one can answer these questions with real accuracy.

However, scientists, insurance companies, and others have been able to estimate some of the costs of air pollution. Following are some examples:

• In California, damage to crops from air pollutants is at least $100 million yearly. The cost of damage to crops nationally is at least five times that much.

• In Syracuse, New York, the furnace in an institution broke down. The result was that about 225 pounds of soot were blown out of the chimney and deposited in the surrounding area. Automobile damage, clothing loss, and cleaning costs came to about $38,000.

• The City Hall of New York was recently repaired at a cost of about $4 million. The cause of the damage? Sulfur dioxide and other pollutants had eroded the building's marble facade.

• A person who owns a house in the suburbs pays from $200 to $300 each year for pollution damage— painting, cleaning, injury to shrubs and trees, etc. These costs are in addition to what would normally be paid out.

• Not all of the coal and oil used as fuel in power plants and factories is completely burned. The unburned and partially burned fuel goes up the chimney and out into the air. Some $300 million worth of fuel is wasted each year in this way. But the costs don't stop there. The damage to the environment costs millions of additional dollars.

POLLUTED AIR IS EXPENSIVE II

• It has been estimated that some $3 billion of gasoline evaporates from American automobiles each year. This would average about $30 annually for each automobile in the country.

• All in all, it is estimated that damage due to air pollution costs about $12 BILLION each year in the United States alone. This averages out to about $65 per person each year.

To Do:

1. The phrases that follow are natural endings to six of the "Polluted air . . ." sentence beginnings at the top of the previous page. Write the proper sentence beginning on the blank lines before the endings. One has already been done to help you get started.

a. *Polluted air causes accidents* by cutting down on visibility.

b. _____
_____
by poisoning the food they eat.

c. _____
_____
causing wires to be exposed.

d. _____
_____
which results in millions of dollars being paid out for cleaning bills.

e. _____
_____
and reduces the number of miles they can be driven.

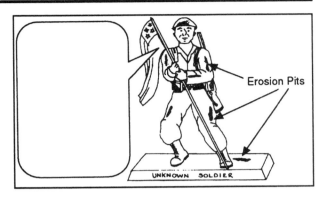

f. _____
_____
and destroys valuable records.

2.a. Circle the phrase at the top of the previous page that is closely related to what is shown in the drawing above.

b. Write these words in the "bubble" in the drawing, "I fought to keep this country free, but how can I fight air pollution?"

3. What is the yearly cost of crop damage in millions of dollar in the United States?

_____

4. Cleaning the exterior of a building might cost as much as $.20 for each square foot. Suppose that each side of a building is 200 feet wide and 300 feet high. How much would it cost to clean all four sides?

_____

5. According to the figures given in the story, about how much does air pollution cost a family of four each year?

_____

6. Underline three phrases at the top of the previous page that are related to vegetation.

A POLLUTED AIR DISASTER

A few years ago, about two weeks before Christmas, in London, England, the fog was so thick that lifetime residents got lost walking through the streets. But it wasn't only the fog that darkened the streets. Smoke from tens of thousands of coal-burning stoves, fireplaces, and factory chimneys mixed with the fog. This mixture of smoke and fog is called **smog**.

The results of the smog conditions were terrible. In the following days, weeks, and months, 3500 more Londoners died than what was normal for the time period. It was the smog that killed them.

The hardest hit were the very old and the very young. People with lung-related diseases were the ones who most often became sick and died. There were many people who did not die immediately but who felt ill for the rest of their lives.

The kind of tragedy that London suffered had happened many times before all over the world, and it is still happening—especially in cities.

Such disasters come about as a result of conditions known as **temperature inversions**. Here is how a temperature inversion works: A layer of warm air forms over an area such as a city. The layer then acts as a kind of lid or "cover" that may not move for days. Meantime, the pollutants released into the air from factories, automobiles, and so on become concentrated below the layer. Sometimes the polluted air cannot even go sideways because the city lies in a valley.

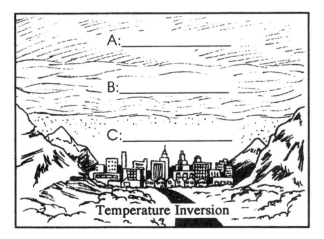

Temperature Inversion

To Do:

1. The diagram shows a cross-section of a temperature inversion. You can add to the diagram by printing:

    Cool Air at A
    Warm Air at B
    Trapped Air at C

2. As shown in the diagram, what is keeping the trapped air from escaping to the sides?

    _____

    _____

3. According to the story, what three things contributed to the smog?

    _____

    _____

    _____

4. What groups of people suffered the most from the smog?

    _____

    _____

5. Circle the sentence in the story that tells you that "smog tragedies" occur frequently.

6. In a temperature inversion, why doesn't the air rising from the earth escape into the atmosphere?

    _____

    _____

    _____

7. Imagine that it is your responsibility to do what is necessary as emergency measures to eliminate a temperature inversion over a city and to help the citizens of the city. List some of the things that you would do.

    _____

    _____

    _____

    _____

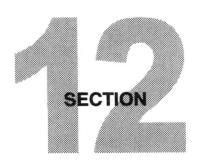

SECTION 12

WILDLIFE IN GEOGRAPHY AND THE ENVIRONMENT

12-1 **Wildlife in the News: Eagles I (Instructor)** 152

12-2 **Wildlife in the News: Eagles II** 153

12-3 **Trees: What We Get From Them I** 154

12-4 **Trees: What We Get From Them II** 155

12-5 **Redwood Trees Make a Comeback I (Instructor)** 156

12-6 **Redwood Trees Make a Comeback II** 157

12-7 **"Space" and How It Affects Animal Populations I (Instructor)** 158

12-8 **"Space" and How It Affects Animal Populations II** 159

12-9 **All About Whales I (Instructor)** 160

12-10 **All About Whales II** 161

12-11 **Wildlife Word Search** 162

WILDLIFE IN THE NEWS: EAGLES I

USA TODAY, a national newspaper, in a recent issue had an interesting headline: "National Symbol Becomes Symbol of Controversy." What is the national symbol, and what is the controversy?

The national symbol referred to is the bald eagle. It appears on innumerable official and unofficial documents. The controversy regards the fact that eagles have been on the endangered species list since the 1940s and are, therefore, protected by strict laws. Now, because their number has increased, there is the possibility that they will be removed from the endangered list and put on the "threatened" list. As *USA TODAY* explained it, "Threatened species cannot be killed or harmed, but development restrictions near their habitats are loosened." Real estate development has the effect of reducing habitat and especially food supplies; reduced habitat may indirectly cause a reduction of eagles. Wildlife protection organizations such as the *National Foundation to Protect America's Eagles* are protesting any change in the eagle's protected status.

Suggested Activity of Completing the Bar Graph

1. Read and/or relate the article. Stress that the eagle is a long-standing symbol of the United States and that its extinction would remove forever the living manifestation of the symbol.

2. Reproduce the facing page and distribute.

3. Direct your students to complete the bar graph on the number of nesting pairs of bald eagles in the lower 48 states, as follows:

a. Write or print the title of the graph on the blank line above the graph: *Nesting Pairs of Bald Eagles in the Lower 48 States*.

b. Enumerate the bold lines on the horizontal axis in intervals of 400: 400, 800, 1200, 1600, 2000, 2400, 2800, 3200, 3600, 4000, 4400, 4800, 5200. 0 has already been printed.

c. At the end of each bar draw a heavy terminal line. Draw diagonal lines in the bar for each year listed on the vertical axis.

4. Table of nesting pairs of bald eagles: (Stress that each pair represents two eagles; thus, 400 pairs represents 800 eagles.)

Year	Number	Year	Number
1963	400	1989	2600
1974	800	1990	3000
1981	1200	1998	5000
1986	1800		

Suggested Activity for Completing the Great Seal of the United States

1. Each part of the Great Seal has a particular significance in the history of our country and relative to the values for which we stand.

2. Direct your students to complete the Great Seal. *Note*: The information in parentheses is for the instructor's use and is not meant to be printed on the seal.

(1) 13 Original Colonies (The 13 stars are breaking through the clouds.)

(2) One Nation Out of Many States (The words on the scroll are written in Latin.)

(3) Bald Eagle (Symbol of self-reliance)

(4) United States Under One Government (The shield indicates protection for all.)

(5) 13 Arrows (We are ready for war, if necessary. Notice arrows in left talon.)

(6) Olive Branch (Notice olive branches are in right talon; we prefer peace.)

Note: Notice how frequently elements of the shield are 13 in number: stripes on shield, leaves on olive branches, olives on branches, stars in clouds.

Name: _____ **Date:** _____

WILDLIFE IN THE NEWS: EAGLES II

Year	Title: _____
1998	
1990	
1989	
1986	
1981	
1974	
1963	

0

Pairs of Nesting Eagles

Source: *USA Today; Wall Street Journal*

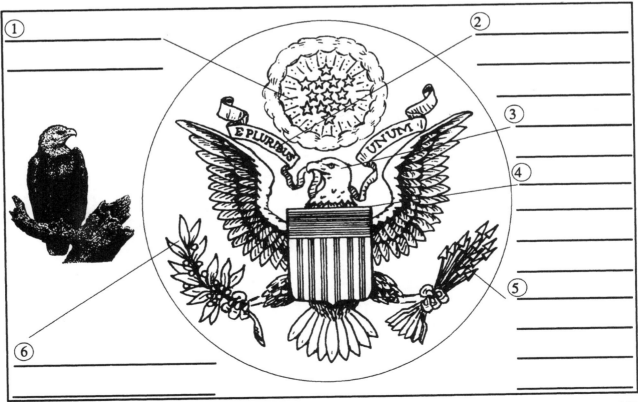

① _____

② _____

③ _____

④ _____

⑤ _____

⑥ _____

153

TREES: WHAT WE GET FROM THEM I

1. About 75 percent of the forests in the United States are commercial forests; that is, privately owned. The remaining forests are owned by federal, state, county, and municipal governments. These forests are important to us because they are sources of oxygen along with other forests throughout the world. Forests are also important to us because they provide the following:

- Watersheds
- Grazing for livestock (cattle, etc.)
- Homes for fish and other wildlife
- Hunting

- Fishing
- Camping
- Hiking
- Lumber and other products

Trees, of course, are the most important part of a forest. It would be impossible to have life as we know it if we didn't have trees. The diagram on the opposite page shows some of the things we get from trees, but there are many more things that aren't listed—there wouldn't be enough room. You can learn more about the products obtained from trees by following the directions below.

Label these things that we get from trees in the numbered boxes:

1. Nuts, Fruit
2. Oils, Extracts, Decorations
3. Varnishes, Soaps, Drugs, Waxes, Turpentine, Crayons, Insecticides, Perfumes, Chewing Gum, Latex (rubber)
4. Sugar, Syrup
5. Tannin, Oils, Dye
6. Charcoal, Rosin, Pine Oil

7. Smoking Pipes, Tea, Oil
8. Paper, Fuel, Charcoal, Plastics, Rayon, Alcohol, Insulation, (and thousands of paper products such as cups, napkins, writing paper, etc.)
9. Poles, Piles, Posts
10. Flooring, Furniture, Shingles, Construction Lumber, Baskets, Plywood, Sawdust, Chips (as bedding for animals)

Our national parks are places where trees are safe. They are carefully protected: Permission must be obtained before any tree in a national park is cut down.

Complete the picture graph that will show and compare the number of visits to the five most visited national parks and recreation areas in a recent year. Let each symbol represent one million visits. Symbol: ☺

Park Visits	
Golden Gate National Park (California)	17 million
National Capital Parks (Washington, D.C.)	9 million
Lake Mead National Recreation Area (Arizona, Nevada)	8 million
Great Smoky Mountains National Park (North Carolina, Tennessee)	8 million
Acadia National Park (Maine)	5 million

THE FIVE MOST VISITED NATIONAL PARKS AND RECREATIONAL AREAS

Parks

Golden Gate	
National Capital	
Lake Mead	
Great Smoky	
Acadia	

☺ = 1 million visits

Name: _____ **Date:** _____

TREES: WHAT WE GET FROM THEM II

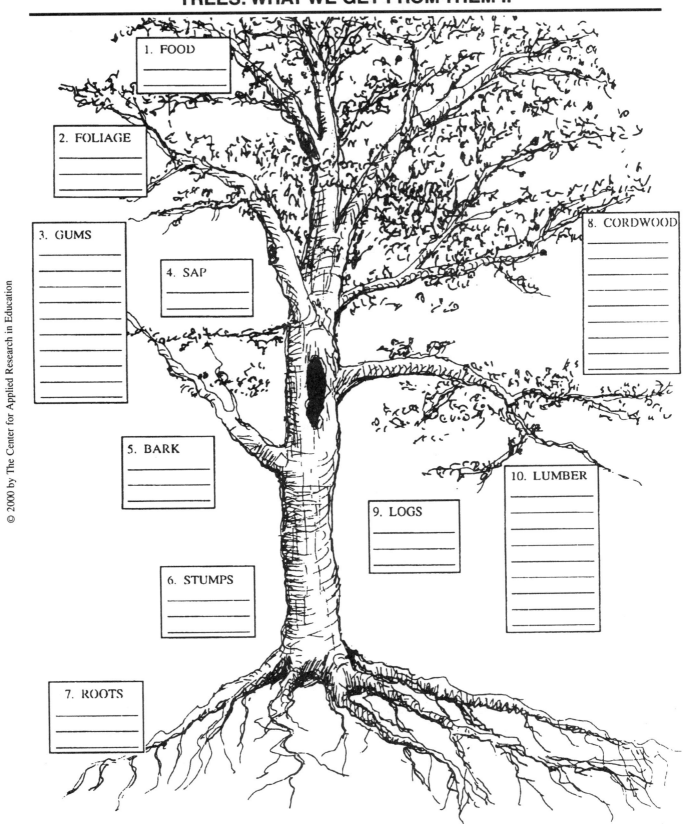

1. FOOD
_____
_____
_____

2. FOLIAGE
_____
_____
_____

3. GUMS
_____
_____
_____
_____
_____
_____
_____
_____
_____

4. SAP
_____
_____

8. CORDWOOD
_____
_____
_____
_____
_____
_____
_____
_____
_____

5. BARK
_____
_____
_____

10. LUMBER
_____
_____
_____
_____
_____
_____

9. LOGS
_____
_____
_____

6. STUMPS
_____
_____
_____

7. ROOTS
_____
_____

REDWOOD TREES MAKE A COMEBACK I

The preservation of the redwood forests of the United States is a great success story. The information on these two pages and the accompanying activities will help your students understand and appreciate these wonderful giants of nature.

The information on this page can be made available to your students orally or by photocopy. The facing page contains a map and other graphics along with questions to which your students can respond.

Background

❑ *Where are redwood forests found?* Redwood forests are found in a narrow band of land no more than 50 miles wide on the Pacific coasts of northern California and southern Oregon.

❑ *Why do redwood trees thrive in these areas?* Because of their immense size, redwoods require vast amounts of water; the trunks of large redwoods may hold up to 8000 gallons. In the areas where they are located, precipitation is very heavy. Fog, which is also common in the region, is thick and frequent. The fog serves to keep the ground wet and the needles of the trees damp.

❑ *How does the root structure of redwoods help them survive?* The roots of redwoods do not penetrate deep into the earth; rather, the roots spread widely from the bases of the trees. This allows for sturdy support when strong winds blow; also, the broad root-spread draws surface water from a wide area.

❑ *How large are redwoods?* One of the trees in Redwood National Park, California, is about 370' tall. If fully laid out on the ground, it would be 70' longer than a football field. Diameters of trunks can be as much as 12'—this would make the circumference more than 38' around.

❑ *What is special about the bark of redwood trees?* The bark is rough and ridged. In a mature tree the bark may be 12" thick. These features make the bark an effective barrier against insect invasions. Another feature of the rough bark is that it is fire resistant. This, plus the fact that fires do not easily sweep through the wet woods means that few trees are lost to fire.

❑ *How long do redwood trees live?* No one can tell how old a redwood tree is until it is cut down. This is because a tree's age is determined by counting the yearly growth rings in the cross-section of its trunk. It is not uncommon for redwoods to be 1500 or more years old. Much of those cut for lumber are about 500 years old.

❑ *Why is the lumber of redwood trees so much in demand?* The reddish hue of the wood makes it attractive; its tannin is a natural preservative; the wood is easily worked; long pieces of lumber, free of knots, may be cut from the trees.

❑ *What are the main uses of redwood lumber?* Although expensive, redwood is especially suitable for interiors of homes, and it makes attractive and lasting siding. Because of its durability and beauty, it is widely used in outdoor furniture. Many house decks are constructed of redwood.

❑ *Are the redwood forests in danger?* In the years before regulation, there was a strong possibility that the redwood forests would be destroyed by commercial lumbering interests. In recent years, however, environmental groups, state and federal governments, individuals, and even lumbering companies themselves, have taken great interest in preserving our redwood forests. Now the cutting of redwood trees is closely regulated. Thousands of redwood seeds are grown in nurseries for replanting. In 1968, Redwood National Park (California), a 110,000 acre preserve, was created. Other national and state parks in the redwood region have been established, and the redwoods in those parks will be protected.

REDWOOD TREES MAKE A COMEBACK II

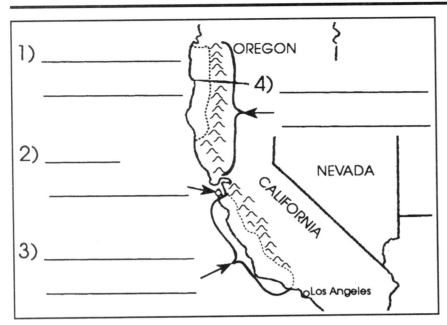

1) _____

_____

2) _____

_____

3) _____

_____

4) _____

3. Why is redwood so popular as an outdoor construction material for things such as siding, decks, and outdoor furniture?

_____
_____
_____
_____

4. What are three things that help redwood trees to survive? Give a brief explanation with each.

a. _____
_____
_____

1.   The areas between the dashed lines and the coast show the redwood forest locations.

a.   Draw three small trees (↑) in each of the forest areas.

b.   Label the following on the map:
(1) Pacific Ocean
(2) San Francisco
(3) Coast Mountain Range
(4) Cascade Mountain Range

2.   When a tree has been cut, the exposed trunk shows growth rings, which can be counted to determine the age of the tree. The thickness of each ring tells the observer if it was a "good" or a "bad" growing year. The thicker the ring, the better the year.

a.   To have an experience in determining the ages of trees, count the rings in the two samples below. Do not count the bark.

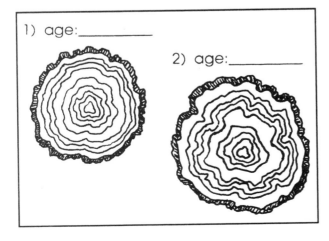

1) age: _____

2) age: _____

b. _____
_____

c. _____
_____

5.   Name four ways that redwood forests are now protected and even growing larger.

a. _____

b. _____

c. _____

d. _____

6.   Underline the following facts told in the story.

a.   Gallons of water a redwood trunk might hold

b.   The thickness the bark of a redwood tree might be

c.   The age of some of the oldest redwood trees

d.   Why forest fires in redwood forests do not burn well

e.   The diameter and circumference of some redwood tree trunks

7.   A three-story building is about 30' high. About how many "stories" high is the 370' high redwood tree mentioned in the story?

_____ stories

"SPACE" AND HOW IT AFFECTS ANIMAL POPULATIONS I

Early people did not require much space for farms, cities, airports, and other such modern facilities. However, throughout the centuries the demands of civilization for space has increased tremendously and is still increasing. Thus, less and less space is left for wildlife. Loss of space, or habitat, has resulted in staggering reductions in both the numbers and kinds of animals. Hundreds, if not thousands of animal species, have become extinct or nearly so.

There are numerous factors that can bring about reductions in animal life:

❑ a decrease in the amount of space available per animal and the amount of healthy food, water, and air the space contains

❑ an increase in the number and kinds of predators in the space

❑ the onset of diseases that may kill the animals

❑ the invasion of the habitat by the same or a different species that has similar survival requirements

❑ the reduction of the habitat by fire, flood, or some other natural disaster such as volcanic eruption

❑ the activities of humans—for example, hunting, fishing, and lumbering

Suggestions for Teaching

You can help your students understand the relationships between space and animal populations by having them participate in the following activity.

a. Photocopy and distribute the "bird space" graphic on the facing page. Explain that the tree provides living space for a colony of birds. Each box represents the amount of space and food one bird needs in order to live for one year.

b. Distribute a handful of split peas, rice grains, candy corn, or some other suitable token to each student to serve as symbols for birds.

c. Direct the students to place one object in any four boxes (4 birds).

d. Explain that after a passage of one year the tree's bird population doubled (8 birds). Four more objects should be placed in four unoccupied squares.

e. After one more year the tree's bird population doubled again, and two birds from another habitat settled in the tree. Place an object in ten more unoccupied squares. How many birds are now in the tree? (18)

f. Additional questions:

❑ What will happen if the additional birds remain? (Each bird will have proportionately less food to eat. There is the possibility that some of the birds will weaken and not survive the winter.)

❑ Suppose the additional birds leave the habitat. Where might they go? (To another habitat—but this might cause problems for the new habitat's original bird inhabitants. There is also the chance that the additional birds might not find another habitat and that, as a result, they might die.)

❑ Suppose a lumbering company decides to cut all the trees. What would happen to all the birds in the trees? (There is a strong possibility that many would die.)

❑ Birds' eggs and young birds are sources of food for some tree-climbing snakes. However, nature balances the number of birds reproduced and the number of birds who do not survive so birds are able to perpetuate their species without becoming too numerous for their habitat. What would happen to the snakes if the birds were forced to leave because of an act of nature or humans? What would happen to animals such as the mongoose for whom snakes are sources of food?

g. Help your students reach the conclusion that all members of a habitat are in some way related and dependent upon one another. If one member of a habitat is increased/decreased, all the members of the habitat will be affected.

Note: An alternative to photocopying and distributing the facing page is to make a transparency of the page and carry out the progression of populations as suggested above.

Name: _____   **Date:** _____

"SPACE" AND HOW IT AFFECTS ANIMAL POPULATIONS II

The Bird Tree

Imagine that this tree is the home for a number of birds. Each birds needs a certain amount of space—and the food it provides—in order to live for one year. Follow your teacher's directions to find out what happens when the birds become too numerous for the space.

ALL ABOUT WHALES I

A recent Associated Press news report contained some discouraging news about whales: Norway announced that it would resume commercial whaling. Equally disturbing to the other nations that are members of the International Whaling Commission (37 nations) was the decision of Iceland, a once important whaling nation, to give up its membership in the Commission. Iceland's decision indicates that it is considering resuming whaling in the future. Finally, Japan, another important whaling nation, let it be known that it was dissatisfied with the prohibition on whale killing.

All the above seems to portend that those whales that were rapidly approaching extinction and granted a reprieve through international agreement—and that have made an excellent recovery—may be in the same position once more. For example, the blue whale, which had numbered some 200,000 before the advent of commercial fishing, had been reduced to only 11,000—5%— by 1985. They have since made a comeback, but they are still in danger of being eliminated from the oceans.

Although the 37 nations that are part of the International Whaling Commission agreed not to hunt whales for commercial purposes, there are some exceptions. The Commission may permit nations to kill a specified number of whales for "scientific research." Also, certain native people who have relied on whales for food and other products for centuries are permitted to catch a limited number of whales. Such whaling is permissible for the natives of Alaska, Siberia, Greenland, and St. Vincent and the Grenadines in the British West Indies.

Following are some interesting facts about whales:

❑ Whales are not fish; they are mammals. They are thought to be among the most intelligent of mammals. Being mammals, the young are born alive and feed on their mothers' milk.

❑ When whales surface, they blow air out of their lungs and then take in more air. In a matter of seconds, the largest whales breathe in as much as 500 gallons of air. Some whales can stay under water for half an hour or more.

❑ Whales, depending on the type, can range from 4 feet in length to 100 feet and can weigh from 50 pounds to 100 tons.

❑ At one time, thousands of years ago, whales were land animals that gradually made the sea their habitat. Through evolution, such appendages as legs have disappeared and been replaced by flippers. Vestiges of the past such as tiny hip bones still remain.

❑ There are at least 75 different kinds of whales divided into two major groups: ***baleen whales***, which do not have teeth, and ***toothed whales***, which do have teeth. There are about ten different kinds of toothed whales. Baleen whales eat mostly plankton (tiny plants and animals), whereas toothed whales eat fish, squid, and cuttlefish.

❑ Historically, what was obtained from whales that made them so commercially desirable and that has led to their near extinction in some cases?

• Whalebones

• Oil for lamps

• Meat for humans and pets

• Oil for margarine, soaps, lubricants, waxes, explosives

It should be noted that the main reason for the renewed desire to hunt for whales today is the meat and the protein they provide. Most other products obtained from whale carcasses now have substitute sources.

❑ Whales, dolphins, and porpoises are part of the marine mammals known as ***Cetacea***. All cetacea are carnivorous to some extent, but only the killer whale eats warm-blooded animals such as seals. Strangely enough, the largest of whales—the blue whale—feeds almost exclusively on the smallest of food; i.e., krill, which is made up of plants and animals that are micro-scopic.

ALL ABOUT WHALES II

Follow the directions below and you will learn about the world's largest animals. Then, with your classmates and friends, you may be able to do something together to help whales.

If you live near the Atlantic or Pacific Ocean it is possible to take a trip on a whale-watching boat and observe whales as they migrate in the changing seasons.

Diagram of a Typical Baleen (no teeth) Whale

On the diagram print the names of the parts as indicated below.

A: **FLUKES** (These "tails" move the whale forward with an up-and-down motion.)

B: **DORSAL FIN** (These help to keep the whale steady in the water—a kind of keel, as in a sail boat.)

C: **FLIPPERS** (These help the whale "steer" and keep balance.)

D: **BLOWHOLES** (Whales inhale and exhale air through these holes. Some whales have two holes; some have only one.)

E: **BALEEN** (These are comblike bones in the whale's mouth that are used to obtain food. Huge quantities of water enter the mouth and are squeezed out. Plankton and small fish remain behind the baleen and are then swallowed.)

F: **PLEATS** (These are folds similar in appearance to an accordion. They expand when a whale takes in food and water.)

G: **EYES** (Whales have eyes on both sides of the head for front and side vision.)

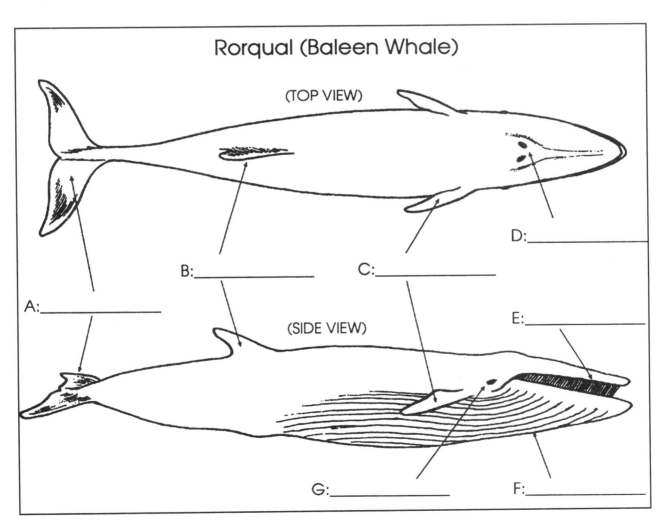

Rorqual (Baleen Whale)

(TOP VIEW)

(SIDE VIEW)

A: _____   B: _____   C: _____   D: _____   E: _____   F: _____   G: _____

Name: _____ Date: _____

WILDLIFE WORD SEARCH

C	A	R	N	I	V	O	R	E	B	D
W	O	L	F	H	S	M	T	N	A	B
F	P	S	O	W	E	N	E	D	C	A
O	R	N	X	H	A	I	E	A	T	T
O	E	A	S	A	L	V	W	N	E	G
D	D	K	T	L	I	O	N	G	R	R
C	A	E	B	E	A	R	Z	E	I	A
H	T	Q	U	L	Z	E	B	R	A	B
A	O	M	I	G	R	A	T	E	E	B
I	R	E	A	G	L	E	H	D	E	I
N	B	T	F	E	X	T	I	N	C	T
H	A	B	I	T	A	T	M	O	L	E
I	S	P	E	C	I	E	S	A	N	O
V	E	R	T	E	B	R	A	T	E	S

© 2000 by The Center for Applied Research in Education

Word Search

The word search puzzle contains 23 words or phrases, all of them related to wildlife in some way. The words are formed across and down. Draw a circle around each of the words. One word has been encircled to help you get started.

Words

bacteria	extinct	mole	species
bat	food chain	omnivore	vertebrates
bear	fox	predator	whale
carnivore	habitat	rabbit	wolf
eagle	lion	seal	zebra
endangered	migrate	snake	

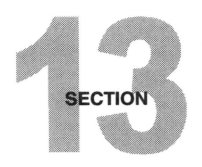

EARTH'S NATURAL WONDERS

13-1   **All About Earthquakes I (Instructor)** 164

13-2   **All About Earthquakes II** 165

13-3   **Volcanoes: Earth's Chimneys (Instructor)** 166

13-4   **The Making of a Volcano** 167

13-5   **All About Glaciers I (Instructor)** 168

13-6   **All About Glaciers II** 169

13-7   **Understanding Icebergs and Ice Sheets** 170

13-8   **Understanding Hurricanes** 171

13-9   **Geysers: Natural Water Fountains** 172

13-10 **Caves: How They Are Made and How They Were Useful in Early Times** 173

ALL ABOUT EARTHQUAKES I

Background

The year 1999 saw one of the most destructive earthquakes in years. It occurred in the country of Turkey. Hundreds of buildings were destroyed, and thousands of people were killed. Some of those who were killed have not yet been found; the materials from the tumbled buildings covered them completely.

Another terrible earthquake occurred years ago in California. In 1906 the city of San Francisco suffered a terrible loss. At least 700 people were killed. As destructive as the Turkey and San Francisco earthquakes were, there have been some that were much worse. For example in Kansu, China, in 1920, an estimated 180,000 people died as the result of an earthquake.

Facts About Earthquakes

• Earthquakes occur for a variety of reasons. One important reason is that Earth's outermost layer is made up of a number of "plates." These plates sometimes move because within the earth there is tremendous heat, as much as 1500°F. Heat rises and tries to escape between and through the plates. Consequently, the plates may be displaced and/or ruptured.

• In their movements, the plates or parts of the plates may drop, rise, slip sideways, or move farther apart. When such movements occur, the surface of the earth is affected. Buildings split and crumble or fall into large cracks; roads, railroad tracks, and bridges are destroyed; the course of rivers and streams may be changed; water pipes break, and water gushes out and causes floods; telephone poles and electric lines are ruined; natural features such as mountains may have avalanches or be split and leveled; and flat land can become bumpy and even hilly or mountainous.

• Earthquakes most frequently occur along the western coasts of North America and South America; the eastern coast of Asia from Japan to the northwest; the coast of the Mediterranean Sea; and then eastward along the northern border of the Arabian Peninsula, India, and the islands northeast of Australia including Indonesia. Almost all the "belts," as the locations are called, are "young" mountain ranges such as the Cascades, Andes, and Himalayas. However, earthquakes also occur in isolated places such as Iceland and New Zealand.

• Many earthquakes take place on ocean bottoms. Because there are no structures there such as buildings and bridges, very little damage occurs. However, they may create huge waves as the ocean bottom upheaves. The waves the underwater upheaval creates can travel as fast as 400 miles per hour. As the waves approach a shore, the bottoms of the waves slow down due to friction, but the tops of the waves hurtle on. Other waves crowd behind the foremost waves. The water piles up and may be as high as 25 feet when it hits the shore. When the *tsunami*, as the Japanese call such waves, hits the land, everything in its path gives way: boardwalks collapse, buildings crumble, and large ships are driven inland. People caught by a tsunami have little chance of surviving.

• When an earthquake causes a gas line to break, the danger of fire is great. The slightest spark can start a fire that will be difficult to stop. The fire spreads quickly, especially in cities where wooden structures have fallen. To render the situation even more difficult, fire fighters may be hindered because streets are blocked, and, if they reach the fire, the water mains may be broken and, therefore, useless.

• It is virtually impossible to accurately predict when an earthquake will occur. However, earthquakes do occur in cycles, but the cycles are highly irregular. The best protection in places where earthquakes are likely to occur is to build structures in such ways that an earthquake will result in minimum damage. For example, reinforced concrete—concrete with imbedded steel cables—will withstand earthquakes better than simple poured concrete with no inner lacing or metal rods. A brick and mortar building offers an earthquake very little resistance. In some earthquake-prone cities, especially in Japan and on the western of the coast United States, earthquake resistant structures are required.

Name: _____ **Date:** _____

ALL ABOUT EARTHQUAKES II

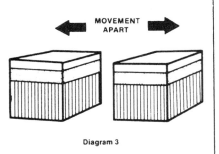

Diagram 1: Cross-section of the Earth

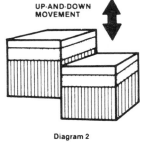

UP-AND-DOWN MOVEMENT

Diagram 2

MOVEMENT APART

Diagram 3

SIDEWAYS MOVEMENT

Diagram 4

Diagram 5a: Before Earthquake   Diagram 5b: During Earthquake   Diagram 5c: After Earthquake

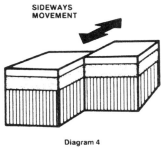

Earthquakes

What makes the earth "quake," or move? Many earth scientists believe that there are a dozen or so *plates* that cover the surface of the earth. If you can imagine a cracked egg, you will have an idea of what the earth looks like with its plates.

The plates, which are 50 to 70 miles thick, move slowly on the *mantle* of the earth. Diagram 1 shows a cross-section of the earth, some of the plates, and the mantle on which the plates "float."

Sometimes one plate will drop lower than the plate next to it (Diagram 2). Sometimes the plates move away from each other (Diagram 3). Sometimes the plates slide against each other (Diagram 4).

When the plates move, the things on the surface of the earth move with them. When that happens, the result may be what you see in Diagrams 5a, 5b, and 5c.

Note: Realize that the diagrams show perfectly square earth parts. In real life they would not be square; they would be irregular. Also, mountains, valleys, streams, etc., would be present.

To Do:

1. Facts from the story titled "Earthquakes":

Number of plates: _____

Thickness of plates: _____

Part of Earth on which plates float:

_____

2. Diagram 1 shows a cross-section of Earth. What are the names of the three main parts?

_____ _____ _____

3.a. Diagram 2 shows only some of the plates. How many are shown? _____

b. Which part of Earth is the thickest?

_____

4. Diagram 2 shows two of the plates moving. What is the direction of the movement?

_____

5. Study Diagram 3. Imagine that a house stood on the place where the two plates moved apart. What would happen to the house?

6. What is the direction of the movement of the plates in Diagram 4?

_____

7.a. Study Diagrams 5a, 5b, and 5c. What kind of plate movement occurred? _____

b. What were the results of the movement after the earthquake? _____

_____

c. What do you think would happen to the barn and house if they stood on the line between the two plates?

_____

_____

8. Which of the movements might result in a trench or depression in the earth?

_____

VOLCANOES: EARTH'S CHIMNEYS

Background

Volcanoes are one of nature's great forces for altering the face of the earth. The materials they spew have created mountains and have filled in valleys and other depressions. Lava from volcanoes has leveled forests and turned the courses of rivers. Islands have been destroyed by volcanoes, and cities have been obliterated.

Following are some more facts about volcanoes:

• Many islands are of volcanic origin, including the Hawaiian Islands. Here is how they were formed: Underwater eruptions gradually build up layers of lava. Eventually, the pile of dried lava emerges from beneath the water and an island is begun. Continued eruptions enlarge the island. Gradually, seeds take hold in the fertile lava, and vegetation grows.

• The world has volcano "belts," where volcanic action is most likely to occur. One belt is the mountain ranges that run along the west coast of North America and South America, from Alaska through Chile. Another belt is located on the east coast of Asia, from Siberia to New Zealand. Volcanoes are especially active in the islands of Southeast Asia.

• In 1883, a volcanic eruption on the island of Krakatau, in Indonesia, resulted in the deaths of 36,000 people from lava, debris, and a tidal wave caused by the explosion.

• About 65 percent of all volcanic eruptions occur in the northern hemisphere.

• The contiguous United States has two active volcanoes: Lassen Peak, California, and Mount St. Helens, Washington. However, Hawaii and Alaska have several volcanoes. There are 32 active vents in the Aleutian Islands. Mauna Loa in Hawaii is the largest volcanic mountain in the world, with a crater of 3.7 square miles.

• Even Antarctica has five active volcanoes, the last two of which were discovered in 1982.

• When volcanic ash falls on land, it eventually becomes part of the soil. It is a natural fertilizer that has significantly enriched soil in many parts of the world.

Volcanoes and Air Pollution

When a volcano erupts, it spews out smoke. The smoke is a mixture of gases. Some of the gases include carbon dioxide, hydrogen sulfide, and sulfur dioxide. The smoke is made black by the fine dust that it carries. The gases and dust not only are injurious to the atmosphere, but also may cause problems among the elderly and people with breathing difficulties.

One of modern history's most significant volcanic eruptions occurred in 1992 in the Philippine Islands close to the United States airbase. The eruption blew an estimated two cubic miles of volcanic ash into the atmosphere. Some of the ash was carried by atmospheric winds for thousands of miles before it settled to earth. So much ash fell on the base that it was closed down and abandoned. Still another negative result was that the vegetation and crops in the near vicinity were buried and lost.

There was a positive result of the Pinatuba eruption, as it was called. The fine volcanic ash that was carried many miles away had the beneficial effect of fertilizing the soil on which it eventually settled. Even the soil that was thickly covered by ash will, in the distant future, be enriched.

Suggestions for Teaching

Prior to assigning the next page, "The Making of a Volcano," it would be helpful to develop interest and knowledge by imparting the information on this page to students. It may be necessary to explain the diagram on the next page as a cross-section that reveals the "workings" of a volcano.

Name: _____   Date: _____

THE MAKING OF A VOLCANO

The material beneath the outer surface of the earth is very hot. It is so hot that rock is melted into a thick molasses-like liquid called *magma* (1). As the magma becomes heated, it expands. Gases and steam form. The pressure against the rock surrounding the magma is tremendous. If a spot in the earth's crust is weak, the gases burst out of that weak spot. Rocks, ashes, sparks, and fire shoot into the air like giant fireworks, and the magma flows out of the opening in the earth.

The "pipe" that the materials use to escape to the earth's surface is called a *conduit* (2). The hole at the earth's surface is a *crater* (3).

The magma that flows into the open air takes a new name—*lava* (4). As the lava flows, it cools and hardens into layers. Over a long period of time, perhaps hundreds or thousands of years, the layers may take the shape of a *cone* (5). The cone may reach thousands of feet into the air.

1. In the story in the box above, words that are special for understanding volcanoes are printed in italics. Each word is followed by a number in parentheses. On the lines in the diagram, write the words that identify the parts of the volcano.

2. According to the diagram, magma and gases escape through the volcano crater. What other means of escape is also used?

_____

_____

_____

3. What causes magma to be formed?

_____

_____

4. What causes the magma to flow out of a weak spot in the earth's surface?

_____

_____

5. Thought question: What are two ways that trees on the sides of the volcanoes could be destroyed?

_____

_____

_____

6. An erupting volcano is spectacular. To make the volcano more realistic, color it as follows:
- Magma: Brownish-orange to show heat
- Cloud: Dull black
- Bright flashes from the crater: Yellow and red
- Sky: Light blue

ALL ABOUT GLACIERS I

Suggestions for Teaching:

Some of the questions on the facing page are based on the information on this page. You may want to use the questions as a study guide; that is, have the students respond to the questions as you present the information. Or, you may choose to photocopy this page. If so, your students can respond to the questions as they read the information.

Background

A glacier is a huge amount of snow and ice that may have taken hundreds of years to form. Snowfall after snowfall fell from the skies and became deeper and deeper. As the snow deepened it became heavier, and the resultant pressure changed the snow into ice.

Glacier Facts

• There are several kinds of glaciers. The largest glacier is known as a *continental glacier*. The reason it has that name is because of its great size. The largest continental glacier actually covers the continent of Antarctica. The continent is completely covered with snow and ice except for its highest peaks. At the South Pole, the Antarctic continental glacier is about one and one-half miles thick. The island of Greenland is another example of the continental-type glacier. The glacier covers more than 600,000 square miles—more than twice the size of Texas. An enormous amount of the world's water, in the form of ice and snow, is held captive in them and other kinds of glaciers.

• *Valley glaciers* are more common than continental glaciers. As the word *valley* indicates, the glacier forms in an area with a relatively flat bottom and slopes on two sides. The valley floor sometimes slopes downward from the head of the valley. Snow and ice pile up in the valley; the glacier is formed, becomes larger and heavier, and begins to move. Two things help it move. One is that the snow and ice become so heavy that the weight "squashes out" the glacier, just as soft clay flattens out when pressed by the heel of a hand. Then, the glacier "slides down" the sloping floor of the valley. The speed at which glaciers move varies greatly. Some may move only a few feet each day, and some may move hundreds of feet daily.

• Valley glaciers have been called "rivers of ice." This is because they flow down a valley as a river might. Moreover, they can be the sources of rivers. As parts of the glacier melt from the sun, the melted water, very cold, gathers together and forms into streams and rivers.

• Glaciers are so powerful that anything in their way that is loose is either pushed before the glacier or taken up into the glacier. A glacier can "sweep a valley clean." There comes a time when a glacier melts backward from the front edge. Ridges are formed from the material deposited at the leading edge of the glacier. These ridges are called *terminal moraines*. Then, as the glacier retreats, the material that it has gathered on its trip down the valley settles on the valley floor. However, those deposits may have traveled hundreds of miles from where they originated. People who see these piles of stone may ask how they got there. The answer is, of course, "The glacier brought them."

• As the glacier moves down the valley, the top of the glacier moves faster than the bottom. The bottom slows down because of the friction created as it travels the rough valley floor, whereas the top of the glacier is much less blocked. It also should be noted that, as the glacier meets the sides of the valley, it picks up debris. Later, when the glacier retreats, the "side debris" is left behind in what are called *lateral moraines*.

• As the glacier leaves the valley down which it has traveled, it spreads out and resembles a snow field. This snow field is called a *piedmont glacier*. Piedmont means at the foot (pied) of the mountain (mont). As might be expected, piedmont glaciers are not as deep as valley glaciers.

ALL ABOUT GLACIERS II

A glacier is formed when many snowfalls fall on a large area over a period of years. The unmelted and piled-up snow comes to have tremendous weight. This weight causes the snow at the bottom of the pile to turn into ice. The newly formed glacier may be hundreds of feet thick. After a while, the glacier moves forward and slowly makes its way down the valley. Such a glacier is called a **valley glacier**. The glacier shown in the drawing is a valley glacier. As you can see, the valley glacier has mountains on both sides.

As the valley glacier emerges from the valley, it spreads out and becomes a **piedmont glacier**. A piedmont glacier is not as thick as a valley glacier. As the piedmont glacier melts, the water that runs off may start rivers and streams.

To Do:

1. Label the glacier drawing as follows.
   Box A: Snow
   Box B: Valley Glacier
   Boxes C and D: Mountains
   Box E: Valley Glacier
   Box F: Piedmont Glacier
Note: Notice that Valley Glacier E is a glacier entering the larger glacier.

2. What causes the fallen snow to turn into ice?

_____

3. Valley glaciers are sometimes called "rivers of ice." Why is this a good way to describe valley glaciers?

_____

4. Ice at the bottom of a glacier is said to be "old ice." Think of a reason why this is true.

_____

5. Circle the sentence in the explanation above that tells how a piedmont glacier is formed.

6. Why do streams and rivers sometimes get their starts from the front of glaciers?

_____

7. When a valley glacier advances, what happens to the stones, gravel, and soil in its path?

_____

8. What name is given to the materials deposited at the front of a glacier after a valley glacier melts back?

_____

9. What is the name given to very large glaciers such as those that cover Antarctica and Greenland?

_____

10. What name is given to the ridge of material that forms on the sides of a valley glacier?

_____

UNDERSTANDING ICEBERGS AND ICE SHEETS

Millions of people have seen the motion picture "Titanic" or have heard or read about it. What was it that caused that great ship, thought to be unsinkable, to sink to the bottom of the sea?

It was an iceberg of great size that caused the ship to sink on its very first voyage from England to New York. The iceberg tore a 300-foot long hole in the ship. Water poured into the ship, and about 1500 people died from drowning and other causes. The tragic event occurred in April 1912 at a remote place in the North Atlantic Ocean, some 1600 miles from its destination.

Some Facts About Icebergs

•   When a glacier moves and finally meets the ocean, large chunks drop off the front of it. The drop-off occurs when the front edge of the glacier is weakened by the warmer ocean water; then, the weight of the ice causes pieces of the front edge to break off. The same kind of break also occurs along the edges of ice sheets, such as those in the Antarctic Ocean off the coast of Antarctica.

•   Icebergs are made of fresh water, so it was not unusual, especially in the days of sailing ships, for a ship to stop near an iceberg and send out small boats to take fresh water from its pools.

•   Icebergs and ice sheets come in all sizes. One ice piece that broke away from an Antarctic ice sheet was much larger than Connecticut. A large iceberg can tower as much as 400 feet above the surface of the ocean.

•   It is easy to see an iceberg above the water, but it is not easy to detect the part of the iceberg that is below the water. Most of an iceberg—some 80% to 90% of its total mass—is below the water's surface. Often, shelves protrude from the underwater parts. It is the underwater parts that ships are most likely to hit.

•   Once icebergs break loose, they float and eventually enter water currents. They may be carried three or four hundred miles before they finally melt away.

•   Icebergs that began as glaciers often carry large amounts of debris such as gravel, soil, sand, and large rocks. This occurs because, as the glacier moves over the land, it collects debris which then ends up in an iceberg. When the iceberg finally melts, the debris falls to the ocean floor.

To Do:

1.   Decide whether the following statements are true (T) or false (F). Draw a circle around the correct answer.

a.   The *Titanic* had sailed many times before striking an iceberg.   T   F

b.   The *Titanic* struck the iceberg in the South Atlantic Ocean.   T   F

c.   The *Titanic* sank a few miles after it left its New York Port.   T   F

d.   One of the reasons ice breaks off from the edge of a glacier is that the warmer ocean water weakens it.   T   F

e.   Water from an iceberg is too salty to drink.   T   F

f.   Pieces of ice that break off from an ice sheet are never larger than a football field.   T   F

g.   The edges of icebergs are steep and do not extend outward under water.   T   F

h.   It is not unusual for icebergs to float hundreds of miles before melting.   T   F

i.   When icebergs melt, they often dump large rocks in the water.   T   F

2.   Circle the sentence that tells how much of an iceberg is below the surface of the water.

3.   Circle the sentences that tell how large an ice piece can be after it has broken away from an ice sheet.

4.   The diagram below shows a typical iceberg. Draw diagonal lines (////) in the part below the surface, and draw horizontal lines (≡≡≡) in the part above the surface.

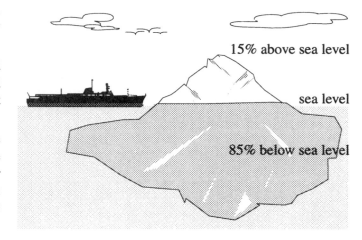

15% above sea level

sea level

85% below sea level

UNDERSTANDING HURRICANES

•   The most damaging hurricane in 1999 carried the name "Floyd." Hurricane Floyd was said by one newspaper, *The Philadelphia Inquirer*, to be "one of the most fearsome storms of the century." The article called the hurricane a "monstrous 400-miles-wide storm." The article also told about the approximate million people who left their homes in the coastal regions of Georgia, South Carolina, and North Carolina to escape the fury of Hurricane Floyd.

•   A hurricane is a tropical storm with winds of 75 miles per hour or more. The winds turn much as a spinning top turns. Hard driving rain, lightning, and thunder are all parts of hurricanes.

•   Hurricanes occur from June to November. September is the month in which most hurricanes happen.

•   At the center of a hurricane's twirling winds is what is known as the *eye*. The eye may be about 20 miles across, and it twirls in the same direction as the winds, but much more slowly—perhaps 20 or 30 miles per hour.

•   The eye of the hurricane is relatively calm, but a sequence in movement often occurs that can be devastating: The outer winds blow fiercely and cause terrible destruction. Then, the eye passes over the area and there is relative calm. However, after the eye passes, more fierce winds follow and cause new destruction. So, the sequence is STORM-CALM-STORM.

•   Hurricanes that originate in the water off the southeast coast of the United States generally move west and northward toward the coast of Florida. However, they may change direction at any time. Also, they may stop traveling and, instead, remain over a particular area for several hours. Then, as hurricanes begin to lose their power, they may veer off the coast and head eastward out to sea, where they gradually die out.

To Do:

1.   To help you understand and remember the facts of the story, circle the words, phrases, or sentences that tell you the following:

a.   the months between which hurricanes occur
b.   the name of 1999's most terrible hurricane
c.   how the newspaper described Hurricane Floyd
d.   the speed at which a hurricane's winds blow
e.   the children's toy that resembles the twirling winds of a hurricane
f.   weather features that are part of a hurricane
g.   where the eye of a hurricane is located
h.   how wide the eye may be
i.   the speed at which the eye twirls
j.   the sequence that some hurricanes follow as they pass over an area
k.   where some hurricanes finally die out

2.   Place a dot at each latitude and longitude listed below. Then, connect the dots and you will have the path of a typical hurricane.

a.   20°N–62½°W (Begin)   e.   27½°N–80°W
b.   22½°N–65°W   f.   30°N–82½°W
c.   25°N–70°W   g.   35°N–75°W
d.   25°N–75°W   h.   40°N–70°W (End)

SOUTHEAST UNITED STATES AND THE WEST INDIES

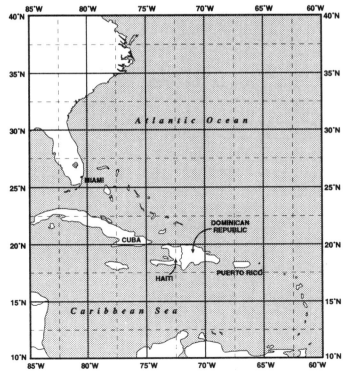

GEYSERS: NATURAL WATER FOUNTAINS

Imagine a column of hot water spouting out of the ground for perhaps ten feet or more than 100 feet, then stopping. In a few minutes, or maybe an hour or even longer, the spouting begins again. This sequence may go on for days, months, or years. What you are imagining, and can even see in certain places in the world, is a *geyser*.

Study the information below and you will better understand one of Earth's most interesting wonders.

• What causes a geyser? Cold water drains into the earth through cracks and holes. As the water drains deeper into the earth, it meets greater and greater heat. Of course, the water becomes hotter and hotter. Finally, the draining hot water collects into a column. The already hot water gets even hotter when it meets hot rocks at the bottom of the column and turns into steam. The steam rises in the column, also called a channel, and pushes the water above it out of the hole. Steam continues to rise and continues to push water. Then, suddenly, the steam and now-very-hot water burst into the air and take on the appearance of a great fountain.

• After the geyser has stopped spouting, some of the water seeps back into the earth, perhaps to be spouted again as it drains to the hot rocks below. The water in some geysers contains mineral matter such as silica and lime, and, after a long period of time, minerals from the geyser's water build up into beautiful forms.

• Where are geysers located? Most of the world's geysers are located in three places that are far apart: The USA's Yellowstone National Park, New Zealand, and Iceland. The Yellowstone geysers are, by far, the most numerous; some 200 geysers are active. The most famous in Yellowstone is "Old Faithful." It erupts, or spouts, for about four minutes every hour. Another geyser in the park, "The Giant," throws water some 175 feet into the air. The water is hot and can scald visitors. Visitors, especially children, are warned not to get too close to the flying "hot drops."

To Do:

1. The "steps" in a geyser's formation are listed below. However, the steps are not in order. Number them in the proper order.

____ Minerals fall to the ground.

____ Hot water collects in a column.

____ Water spouts.

____ Cold water drains into cracks and holes.

____ Steam rises and pushes water to the surface.

2. The picture below shows the Lone Star Geyser in Yellowstone National Park. Circle the words in the story that tells how the formation at the base of the geyser came into being.

3. Where are the world's three most important geyser locations?

_____ .

_____ ,

and _____

4. Why must people be careful when they visit geysers?

_____

_____

_____

_____

5. To make the Lone Star Geyser more realistic draw small dots (water droplets) in the spray coming out of its center.

U. S. Geological Survey

*The **Lone Star Geyser** in Yellowstone National Park shoots water 25 feet into the air.*

CAVES: HOW THEY ARE MADE AND HOW THEY WERE USEFUL IN EARLY TIMES

Caverns, or caves, were probably the first enclosed places that the earliest human beings occupied. It was much safer to live in a cave than to live in forests or open plains where wild animals could attack. Caverns have entrances that can be blocked and easily guarded. A cave can be warmed by a fire and can serve as a cool, safe place in which to store food. And, of course, there is protection from the weather that brings rain, snow, or high winds.

Here are several different kinds of caves and the ways they are formed.

1. ***Mountain/hillside caves*** are usually formed by water trickling through soft limestone. Over many years, the moving water wears away or dissolves the limestone. Water continues to trickle, and eventually rooms and passages are formed.

2. ***Lava caves*** are the result of volcanic action. A volcano erupts, and hot, thick lava pours down the mountainside. The outer flow of the lava hardens, but the inner flow remains liquid. The liquid lava gradually flows away, and the result is a tunnel-like cave.

3. ***Sea caves*** are formed by the action of waves pounding on rocky and cliff-like shores. Gradually the rocks wear away, and a large hole remains. The sea cave would be difficult to live in because, even though it may be empty of sea water at low tide, it could fill up with water at high tide.

4. ***Waterfall caves*** are formed when tumbling, swishing, and swirling water gradually hollows out a part of the cliff over which water is falling. People sometimes climb up to such caves, enter them, and then watch the water come pouring down.

Caves come in small and large sizes. The Carlsbad Caverns in New Mexico are exceptionally large. The caverns have some very large rooms. The room known as the "Big Room" is more than three-quarters of a mile long, more than 600 feet wide, and almost 300 feet high. There are miles of passages in the caverns; some of them have not yet been explored. There are passages on different levels; an elevator takes visitors to the various levels.

Caves often have strange mineral formations. The formations are made from dripping ceiling water that contains minerals. When the water drops and splashes, the minerals remain on the floor of the cave. As more drips come over a long period of time, a kind of cone is formed that points up to the ceiling.

Also, some of the minerals stick to the ceiling; they gradually begin to look like an icicle. So, the mineral cone on the floor and the mineral cone on the ceiling are pointing toward each other. As time goes on, the two cones will probably meet, and then a ***column*** is formed. The cones from the ceiling *down* are called ***stalactites***, and the cones from the floor *up* are called ***stalagmites***.

To Do:

1. List four advantages that caves gave early dwellers, as compared to living out in the open.

a. _____

b. _____

c. _____

d. _____

2. How are mountain/hillside caves formed?

_____

_____

3. What shape do lava caves often take?

_____

4. How are waterfall caves formed?

_____

_____

5. Underline the sentence that tells why sea caves may not be good living places.

6. Circle the part of the sentence that tells the size of Carlsbad Caverns' "Big Room."

7. In the drawing below, draw slanted (/////) lines in all the stalactites, horizontal (≣) lines in the stalagmites, and dots (⦂•⦂) in the column.

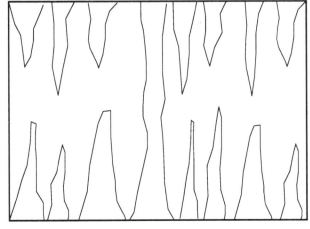

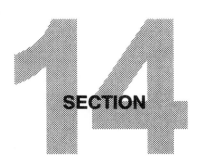

LEARNING FROM PICTURES

14-1 Developing Picture-Reading Skills (Instructor) 176

14-2 Working Together in Switzerland 177

14-3 In the Netherlands Dikes Hold Back the Ocean 178

14-4 A Mexican Village 179

14-5 Animal Friends and Enemies in Australia 180

14-6 Life in a Desert 181

14-7 Fishing in the United States 182

14-8 Life on the Plains in Early Times 183

14-9 Nature and Humans Change the Face of Earth 184

DEVELOPING PICTURE-READING SKILLS

What are the kinds of things students should look for when they are examining pictures from a social studies point-of-view? Are there basic questions that can be asked about a picture that will yield most of the facts and inferences it holds?

The answers to these questions can be found in a definition of social studies: those studies concerned with the relationships of humans to humans and of humans to their environment. When studying a picture it is helpful to keep these two relationships in mind.

Not all students will have the ability to derive facts and inferences from pictures. Some students need specific guidance "through" a picture. Furthermore, not all students will derive the same facts and inferences from the same picture or pictures. This is because the same picture will trigger different reactions and associations in different viewers. The kinds and extent of associations students make are dependent upon the knowledge, background of experience, interests, and imagination they bring to the picture.

The following questions should prove helpful in guiding a variety of students from a variety of backgrounds to study pictures. They are designed to help students retrieve social studies information from pictures.

Specific Questions

1. What does this picture tell about the natural environment of humans, including

- vegetation?
- climate?
- landforms?
- weather?
- water forms?
- wildlife?

2. What does this picture tell about the way that humans have adapted to the environment as evidenced by the

- changes made in the natural environment?
- changes in the land and water forms?
- crops grown?
- animals raised?
- clothing worn?
- presence of roads, bridges, and tunnels?
- presence of homes and other buildings?
- tools used?
- ways natural resources are used or misused?

3. What does this picture tell about the lives of humans, including how they

- communicate with other humans?
- provide food for themselves?
- live with other humans in groups?
- entertain and amuse themselves?
- govern themselves?
- transport themselves?
- provide for safety, comfort, health, and convenience?
- work and the conditions under which they work?
- educate themselves?
- are different from, or similar to, humans of other times and places?

Following are some additional elements that are concerned with picture-reading skills.

❑ Think of pictures as containing main ideas and supporting details. However, unlike a typical printed paragraph, a picture may contain more than one main idea.

❑ Picture reading requires more insightful thought than reading words. Words and sentences may do the "thinking" for learners, whereas pictures require learners to do the thinking. The result is that what is learned from pictures may be more meaningful and retained longer. This, along with the visual impact that pictures have, makes them a powerful learning tool.

❑ Facts and details obtained from pictures may be combined to make inferences. For example, a picture that shows crops growing in a desert-like environment invites one to make the inference that water for the crops must be obtained in contrived ways such as dug wells or irrigation ditches. If there are irrigation ditches, the next inference might be that the water comes from the snow-topped mountains in the background.

Suggestions for Teaching

The activity on the facing page is designed to help learners become more aware of details in an illustration. Also, by searching the picture numerous times, the reader can gain a lasting visual impression.

Question 2 is concerned with making inferences from the facts/details in the picture. Answers will probably vary; however, that will make for good discussion.

Name: _____ Date: _____

WORKING TOGETHER IN SWITZERLAND

The picture shows a mountain scene in Switzerland. The time of the year is late summer. The cattle have been taken up to the mountains for pasture. They will return to the valley in the autumn.

To Do:

1. Put a check before all the things listed below that can be seen in the picture.

a. _____ Man splitting logs

b. _____ Eight cows

c. _____ Village in the valley

d. _____ Man carrying hay

e. _____ Young boy with hat

f. _____ Carts and horses

g. _____ Steep trail

h. _____ High mountains

i. _____ Tree-covered mountains

j. _____ Cow being milked

k. _____ Steep cliff

l. _____ Rail fence

m. _____ Forests

n. _____ Rocky land

o. _____ Axe

p. _____ Man wearing suspenders

2. Why is it likely that

a. the man is carrying cheese on his back?

_____

_____

b. the scene is one of late summer rather than early spring?

_____

_____

c. no field crops such as wheat, corn, or oats are grown in the area?

_____

_____

d. the boy is waving good-bye to the men carrying loads down the mountain?

_____

_____

e. all the workers do not return to the village at the end of a work day?

_____

_____

IN THE NETHERLANDS DIKES HOLD BACK THE OCEAN

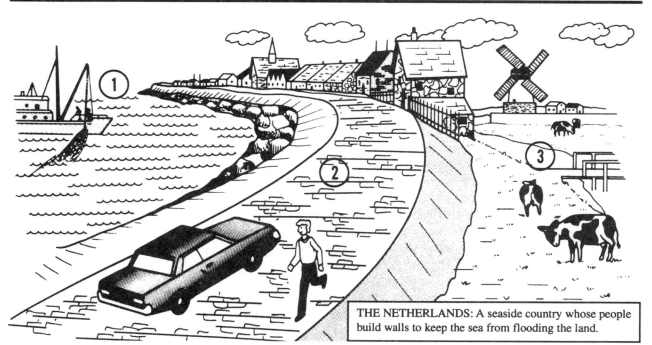

THE NETHERLANDS: A seaside country whose people build walls to keep the sea from flooding the land.

In the picture, the water at ① on the left is the ocean. The ridge of land at ② is a dike. The houses and fields at ③ are below the level of the ocean water. *Note*: Questions 1-8 are thought questions. It will be up to you to think of reasonable answers.

1. What would happen to the land and houses if the dike broke and an opening was created?

_____

_____

2. What are two ways the dike is useful?

a. _____

b. _____

3. What is the purpose of the net on the boat?

_____

_____

4. What two kinds of food can be obtained from the cattle?

a. _____

b. _____

5. Why are there great rocks on the ocean side of the dike, but not on the land side? _____

_____

6. All the houses have steep roofs. Why are they made that way? _____

_____

7. How would the windmill be helpful if the land became flooded with water? _____

_____

8. The land you see was once under salty-ocean water. What had to be done to the land before it became fertile enough to grow crops? Try to think of two things.

a. _____

b. _____

9. How is the automobile's position on the road different than it would be in the United States?

_____

10. Find and list two details about each of the following.

Dike: _____

Houses: _____

Windmill: _____

A MEXICAN VILLAGE

One of the world's largest cities is Mexico City, capital of Mexico. There are more than 16 million people who live and work there. There are also many modern buildings and a fine transportation system. However, many of Mexico's communities are small villages that have not changed much over the years. The picture above shows such a village.

The activity that follows will help you learn some Spanish words, and you will also gain an understanding of Mexican village life.

To Do:

1. Each English word in the box that follows means the same thing as one of the Spanish words in the picture. In the next column, on the line after each Spanish word, write the English word with the same meaning.

blacksmith	grass	road
cactus	hat	roof
carpenter	horse	sky
cart	house	trees
cattle	man	wall
children	mountain	wheel
church	people	woman

2. Circle the following picture details.

- sleeping man
- jar on woman's head
- church cross
- smoke from chimney
- three children
- man with a stick

SPANISH	ENGLISH
sombrero	
caballo	
mujer	
cacto	
hombre	
montaña	
casa	
cielo	
gente	
carro	
carpintero	
ganado	
rueda	
templo	
niños	
tejado	
pared	
camino	
árboles	
hierba	
herrero	

Name: _____   Date: _____

ANIMAL FRIENDS AND ENEMIES IN AUSTRALIA

The four pictures above show a part of Australia that has very little rainfall. This makes it hard to grow crops. However, large herds of sheep and cattle are raised there. The water they need flows under ground from mountains in the east. Then, wells are dug to the water, and the water is pumped out and stored in ponds and tanks.

1.   Each sentence below tells something about one of the pictures. Read each sentence. On the line before it, write the number of the picture it tells about.

_____ Rabbits, kangaroos, and sheep all live on grass.

_____ The sheds in the background will be used in shearing sheep of their wool.

_____ Much needed water is pumped into tanks and ponds.

_____ Wild, doglike animals attack and kill sheep.

_____ Herders, horses, and dogs work together.

_____ Kangaroos can easily jump the fence.

_____ Sheep are almost helpless in any kind of a fight.

_____ Pumps powered by the wind are used where there is no electricity.

_____ A few trees grow in the grass country.

_____ Cattle like these are used for beef rather than milk.

2.   Some "Why" and "How" questions:

a.   Why is the dog in picture 1 chasing the sheep?

_____

_____

b.   Notice the rabbits in picture 2. Why do Australian farmers consider rabbits to be harmful pests?

_____

_____

c.   Notice the one sheep behind the horse and rider in picture 1. He is straying from the herd. But, the dog is busy with another sheep. How will the sheep be put back into the herd?

_____

_____

_____

LIFE IN A DESERT

The picture shows people working together in the world's greatest desert—the Sahara Desert in Northern Africa.

To Do:

1. Put a check before all the details listed below that you see in the desert.

_____ goats   _____ clumps of grass

_____ three camels   _____ tent

_____ a horse   _____ loose clothing

_____ a cart   _____ jugs

_____ rugs   _____ mallet (hammer)

_____ poles   _____ trees

2. Find four other details, and briefly list them on the lines below.

a. _____

b. _____

c. _____

d. _____

3. Check those statements that follow that you think are probably true. Have reasons ready to explain your answer.

a. _____ The people drink goat's milk.

b. _____ The people are moving out of the area.

c. _____ It is easy for animals to find plenty to eat in the area.

d. _____ Children go to school each day.

e. _____ The people sleep on rugs.

f. _____ The people do not eat much fruit and vegetables.

g. _____ When the grass is all eaten by the animals, the people move to another place in the desert.

h. _____ The poles used for the tent came from other places.

i. _____ It is necessary for the people to carry water when they move from place to place.

4. A paragraph has a main idea, and so does a picture. In fact, it's possible that a picture could show more than one main idea. On the lines below write what you think is a main idea of the picture.

_____

_____

_____

_____

Name: _____ **Date:** _____

FISHING IN THE UNITED STATES

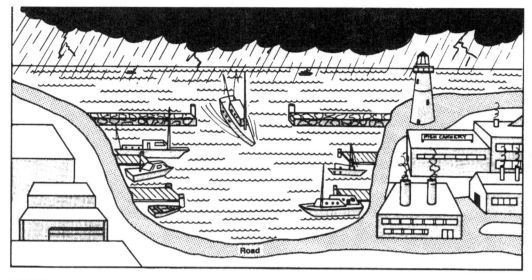

The picture at the top shows a typical fishing port that could be in the United States and other places around the world.

The three numbered pictures show some of the activities that take place after fishing boats return to port.

Note: *This activity will require you to think about what you see in the pictures. Think carefully before you write your answers.*

To Do:

1. What is the purpose of the rock walls at the entrance of the harbor?

_____

2. What is in the picture that will protect fishing boats if they return at night?

_____

3. How can you tell that a storm may soon reach the harbor?

_____

4. How many docks are shown in the picture?

_____

5. Why are the fish cannery buildings so close to the docks?

_____

6. How can you tell that the fish cannery is open and ready for work?

_____

Bottom Pictures

1. On the line below each numbered picture, write a two-word title that tells you what is taking place.

① _____

② _____

③ _____

2. Picture 3 shows fish being placed in packages. What would be two more steps before a package of fish is bought by a customer in a store?

a. _____

b. _____

Name: _____   **Date:** _____

LIFE ON THE PLAINS IN EARLY TIMES

Those pioneers who settled the Great Plains area of the United States had many obstacles to overcome. The absence of wood, scarcity of water, natural disasters such as hurricanes and blizzards, and long distances between homesteads were just a few problems that they faced.

To Do:

1. Each phrase applies to one of the pictures. Write the number of the phrase in the circle in the correct picture.

(1) Traveling by wagon _____

(2) Building a house _____

(3) Digging for water _____

(4) Farming with a plow _____

(5) Trying to stop a fire _____

(6) Clouds of grass-
     hoppers _____

(7) Fighting a blizzard _____

(8) Fleeing a tornado _____

(9) Erecting a fence _____

(10) Stuck on the road _____

2. Each phrase above is followed by a line. One of the words below could be inserted before the last word in one of the phrases to make it more meaningful. Write the letter of the correct word on the line after each phrase.

a frightening
b steel
c muddy
d grass
e Conestoga
f sod
g fresh
h blinding
i barbed-wire
j hungry

NATURE AND HUMANS CHANGE THE FACE OF EARTH

1. The drawings show some of the ways humans and nature change the earth. At the bottom of each picture print NATURE if a natural change is being made. Print HUMAN if a human-made change is taking place.

2. Which picture shows each of the following?

_____   a housing development across the river from a city

_____   stones from a quarry being loaded

_____   a river building a delta

_____   waves carving a hole in a rock

_____   workers landscaping a park

_____   a volcano erupting

3. Which pictures show something that will pollute the air? _____

4. Which picture shows human actions that will improve the environment? _____

5. The "islands of soil" shown in picture 1 have not always been there. After studying the picture, try to tell where the soil came from and how it got there.

_____

_____

6. Try to explain how the waves shown in picture 4 carved a hole in the rock.

_____

_____

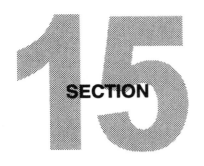

SECTION 15

EARTH, SUN, AND MOON RELATIONSHIPS

15-1 Day and Night: What Causes Them? 186

15-2 The Moon: Some Factual Information (Instructor) 187

15-3 Remembering Moon Facts 188

15-4 The Sun and Its Relationship to Earth (Instructor) 189

15-5 Seasons North and South of the Equator 190

15-6 All About Compasses and Magnetic North (Instructor) 191

15-7 Understanding the Compass: Direction and Time 192

Name: _____ Date: _____

DAY AND NIGHT: WHAT CAUSES THEM?

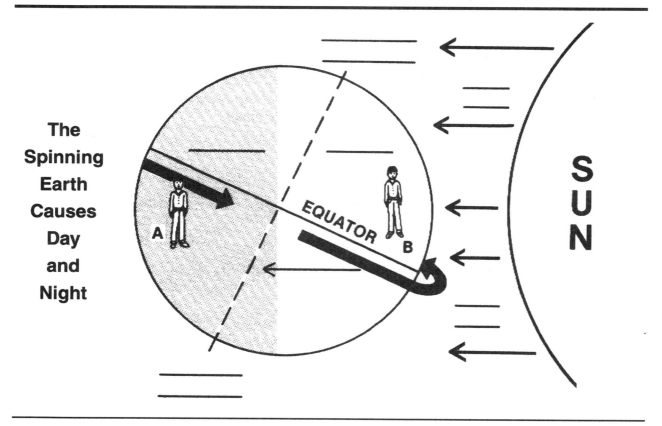

The Spinning Earth Causes Day and Night

Many years ago people thought that the sun rises in the east. But, after considerable study, it was determined that the sun does not "rise." Rather, Earth rotates (turns) on its axis—much as a top turns—from *west to east*. As Earth rotates, half of Earth is getting light from the sun and the other half is experiencing darkness.

In the United States, the Atlantic coast regions see the sun first. People on the other side of the United States (west) are still in darkness. So, as Earth keeps turning new parts of our country see the sun. People on the east coast see the sun about three hours before people on the west coast see the sun.

To Do:

As you can see in the diagram, boy B is in sunlight and boy A is in darkness. But as Earth rotates from west to east, boy A will be in the sunlight and boy B will be in darkness.

1. To make the diagram more interesting and informative, do the following.

a. On the line over the head of boy B, print DAY.

b. On the line over the head of boy A, print NIGHT.

c. At the top of the slanted dashed line print NORTH POLE. At the bottom of the dashed line print SOUTH POLE.

d. On the heavy arrow that is pointing in a west-to-east direction print WEST-TO-EAST. The arrow shows the direction in which Earth rotates.

e. On the arrow pointing to the dashed line print AXIS. That's the axis (axle) on which Earth turns. It takes 24 hours to make a complete turn.

f. Notice the arrows from the sun pointing to Earth. Print SUN RAYS between the two top arrows and the two bottom arrows. Then, lightly color the sun's rays and the eastern half of the diagram yellow.

THE MOON: SOME FACTUAL INFORMATION

Suggestions for Teaching

The information below can be read to your students. Following that, administer the true-false quiz printed on the facing page. Following the quiz, discuss the items and the answers. In this way students will have had three exposures to the moon information—the reading, the quiz, and the review of the quiz. Finally, it would be helpful to give students a photocopy of this moon information page. They can use it for further study—perhaps for a "real" test.

Facts About the Moon

There is no space object closer to Earth than the moon. Although this is true, it must be realized that the average distance of the moon from Earth is 240,000 miles. The word "average" is used because as the moon circles Earth, it is sometimes closer and sometimes farther away. In other words, the moon's orbit is not an exact circle; its orbit is oval-shaped.

• It takes the moon almost 28 days to complete its trip around Earth. In order to make the trip in that time, the moon moves at a speed of about 2300 miles per hour.

• One of the most important effects of the moon on Earth is that it causes tides in the ocean. This is because the moon exerts a "pull" on the part of Earth that is facing the moon. The pull is known as *gravity*. As a result, the pulled water bulges out toward the moon. At the same time, ocean water on the opposite side of Earth is also being pulled and is bulging out. The moon's pull causes *high tide*.

The bulge comes about because water is being drawn from other parts of the ocean. The places where water is being drawn away are experiencing *low tide*. The tides change from low to high about every 12 hours.

• On a clear night the moon is very bright. This may lead to the thought that the moon is a source of light. Rather, the moon's brightness is caused by the sun which shines on the moon; the moon reflects that light.

• There is no water on the moon; the surface is absolutely dry.

• There is no atmosphere on the moon. Consequently, there is no air and there is no sound.

• The pull of gravity on the moon is so slight that a good high-jumper could clear a bar 25 feet high. The jumper would not be injured when he came down because his weight would be less than 20% of what it would be on Earth. In fact, the moon's relatively weak gravitational pull caused the first astronauts who walked on the moon to wear heavy clothing and weights on their shoes to keep them from bouncing up as they walked across the moon's surface.

• During the day, temperatures on the moon climb as high as 220°F. During the night, temperatures fall as low as -95°F.

• There are mountains, plains, and *craters* on the moon. A crater is a deep depression in the surface. There is no vegetation—no grass, trees, brush, or flowers. There is no form of life on the moon. The surface is covered by a fine dust. The dust is not disturbed by winds because there is no wind on the moon.

• In its revolutions around Earth, the moon sometimes comes between Earth and the sun. When this occurs, the sun cannot be seen from Earth. This occurrence is known as a solar eclipse. To *eclipse* means to shut out. So, during a solar (sun) eclipse, the moon "shuts out," or hides, the sun.

Name: _____ Date: _____

REMEMBERING MOON FACTS

Testing Your Knowledge of the Moon

Each sentence below is either true (T) or false (F). Circle T if the sentence is true; circle F if the sentence is false. For a sentence to be true it must be totally true. A sentence that is false might have some true parts; however, if the sentence has any false parts it should be marked false.

1. The sun is closer to Earth than the moon is.   T   F

2. The moon is about 150,000 miles from Earth.   T   F

3. The moon circles Earth in an oval-shaped orbit.   T   F

4. Every 14 days the moon circles Earth.   T   F

5. The moon circles Earth at a speed of about 2300 miles per hour.   T   F

6. The sun is the main cause of tides in Earth's ocean waters.   T   F

7. A high tide is caused in ocean waters when the moon's gravity "pulls" water
    away from another part of the ocean.   T   F

8. People walking on the moon need heavy shoes to prevent bouncing as they walk.   T   F

9. The moon would not be bright if it didn't receive light from the sun.   T   F

10. Astronauts who landed on the moon had to wade through water.   T   F

11. Sounds on the moon are very loud.   T   F

12. Astronauts who landed on the moon had to carry their own air supply.   T   F

13. The weight of an object on the moon is greater than it is on Earth.   T   F

14. Winds on the moon are strong and constantly blowing.   T   F

15. The moon is free of all dust.   T   F

16. Temperatures on the moon are about the same day and night.   T   F

17. There are mountains on the moon but no plains.   T   F

18. The astronauts were able to bring flowers from the moon back to Earth.   T   F

19. The moon has great depressions called craters.   T   F

20. A solar eclipse occurs when the moon comes between Earth and the sun.   T   F

THE SUN AND ITS RELATIONSHIP TO EARTH

Background

Students find it difficult to visualize the motion of the earth in relation to the sun. To help them understand the relationships, it is helpful to use three-dimensional models before reviewing the more abstract diagram.

Following are some suggestions for teaching earth-to-sun relationships.

1. The earth spins on its axis, and the axis is turned at a 23½° angle. To demonstrate this fact, obtain a globe that can be adjusted to a 23½° angle. Show the globe, and then move from the 3-D globe to a more abstract diagram by drawing a circle on the board. To show the incline, draw the Earth's axis 23½° to the right of straight up and down. Also, to fix in the minds of students that the earth is inclined, keep the globe inclined as it stands on a shelf or windowsill in the classroom.

2. The constant incline of the earth is the most significant fact in understanding the seasons of the year. If there were no incline, there would be no seasons because the sun's rays would always strike the various points of Earth with the same intensity

The incline coupled with the fact that the earth revolves around the sun once every year causes the sun's rays to strike the northern hemisphere directly for six months and then strike the southern hemisphere directly for six months. (The drawing that follows shows this quite clearly. It would be helpful to project the drawing via a transparency, accompanied by explanations.)

3. The diagram shows the position of the earth in relation to the sun in the four seasons of the year. When the northern hemisphere is tilted toward the sun, the Arctic has six months of light, and the Antarctic has six months of darkness. The reverse is true when the northern hemisphere is tilted away from the sun.

4. To demonstrate the effect of the incline in a striking way, shine a flashlight directly on the middle of an inclined globe. Students should notice that when the northern part of the globe faces the light, the northern parts are quite light and the southern parts are quite dark. Now, turn the globe so that the southern hemisphere faces the flashlight. The northern parts will be in the dark, and the southern parts will be in the light.

5. The following are facts about the sun.

• Our sun is really a star among several million stars. However, it is the star closest to Earth, and that is why it is so noticeable to people on Earth.

• The sun is some 93,000,000 miles from Earth.

• The sun is very large compared to Earth. The sun's diameter is about 865,000 miles while Earth's diameter is about 8,000 miles. That makes the sun's diameter some 108 times longer than Earth's.

• It takes Earth about 365 days to orbit the sun. This means that Earth is orbiting at a speed of more than 60,000 miles per hour.

• We think it is very hot when the temperature reaches 100°F. However, the sun's surface temperature is about 11,000°F.

• Unlike Earth, the sun is not solid; rather, it is a ball of several hot gases, including helium and hydrogen.

• Earth is only one of nine planets that revolve around the sun. In order of increasing distance from the sun, the planets are Mercury, Venus, Earth, Mars, Jupiter, Saturn, Uranus, Neptune, Pluto. Jupiter is the largest planet, Mercury is the smallest, and Earth is the third smallest.

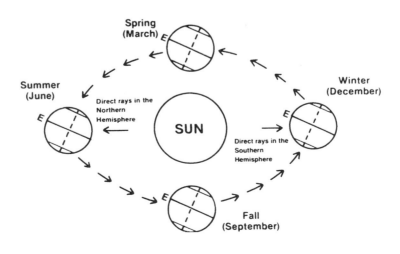

SEASONS NORTH AND SOUTH OF THE EQUATOR

Study the diagram on this page. Then circle the words that correctly complete items 1-10.

1. The globe at the left shows the position of the earth in relation to the sun on (June 22, December 22).

2. On June 22 the (South, North) Pole is inclined toward the sun.

3. On June 22 the direct rays of the sun are at the (Tropic of Capricorn, Tropic of Cancer).

4. On June 22 the sun's rays at the Arctic Circle are (hotter, cooler) than those at the Antarctic Circle.

5. At the Antarctic Circle, the warmer day would be (December 22, June 22).

6. If you were at the place marked G on December 22, you could expect to be (hot, cold).

7. On June 22 the sun's rays strike the (North, South) Pole.

8. On December 22 you could not see the sun at the (North Pole, South Pole).

9. At H on December 22 it is the (warm, cool) season. At E it is (cool, warm).

10. You now know that when places north of the equator are having winter, places south of the equator are having summer. Therefore, when it is spring at A, it must be fall at (B, D).

11. Why would there be no seasons if the earth were not tilted?

_____

_____

_____

_____

_____

EARTH'S POSITION IN RELATION TO THE SUN ON TWO DAYS

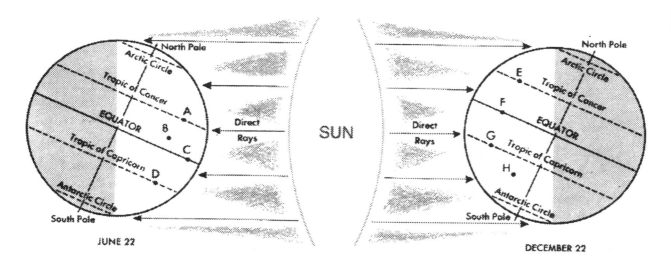

JUNE 22   DECEMBER 22

ALL ABOUT COMPASSES AND MAGNETIC NORTH

Background

Note: Students should find the following information interesting.

In today's world it is possible to go to almost any destination by motor vehicle. But, imagine that, after the road and the road map took you to a place, it was necessary to find your way to a cabin deep in the woods and mountains. And further imagine that there were no trails to follow that would take you to where you wanted to go. If you simply took off into the woods and the place was three or four miles away, there would be a strong possibility you'd never find the place. Even worse, you might get lost.

If you had a compass with you, and you knew how to use it, you could easily find your way to your destination. However, there are some things you would need to know before you could use a compass effectively.

1.   A compass needle points to magnetic north; it does not point to the North Pole.

2.   *Magnetic north* is located some 1200 miles south of *true north*. The vicinity of Prince of Wales Island, about 1200 miles south of true north, is the basic location of magnetic north.

3.   Why is magnetic north magnetic? About 70 miles below the surface of the earth there is a great lode, or deposit, of magnetic materials that has the power to attract a magnetic needle if the needle is suspended on an upright point within a compass housing. The magnetic field is so powerful that it will attract a magnetic needle even if the needle is thousands of miles from the magnetic field.

Figure 1 in the next column shows a magnetic compass with true north and true south labeled. Magnetic north is also shown. Notice how the compass magnetic needle is pointing to magnetic north. There is a difference, called a declination, of 20° between true and magnetic north.

Figure 2 also shows a magnetic declination of 20° on a map of North America. The beginning point is in the state of Maine. Also notice magnetic north is located in the vicinity of Prince of Wales island.

Note: If you are interested in teaching students about compasses, how to set them for cross-country hikes,

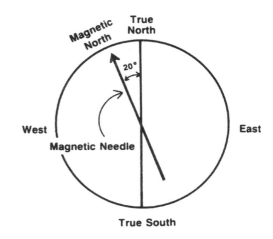

MAGNETIC COMPASS

Figure 1

Figure 2

and how to conduct such hikes, the following book would be of great value.

Be Expert With Map And Compass —
The Orienteering Handbook

Published by:
Silva, Inc., and American Orienteering Services
Highway 19
La Porte, Indiana 46350

UNDERSTANDING THE COMPASS: DIRECTION AND TIME

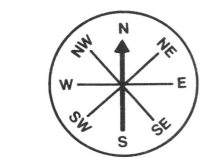

Left Hand   **Right Hand**

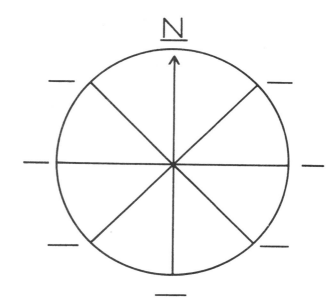

1. As you can see in the drawing, the girl is facing north. You know she is facing north because she is facing in the same direction as the arrow in the compass.

a. In what direction would the girl be facing if she were to turn halfway around? _____

b. Toward what direction is the girl's right hand pointing? _____

c. Imagine that the girl turned so that she was facing northeast. What direction would be directly behind her? _____

d. Imagine that the girl turned and faced in the direction of southeast. What direction would be directly behind her? _____

2. The compass pictured in the next column has no directions labeled on it. Without looking at the compass at the top of the page, write all the directions on the lines. Use abbreviations. After you have completed the lettering, check to see if you are correct.

3. Imagine that you are going to take a compass walk. You have set your compass, and you know in which direction to walk. However, you should have a good idea as to how long it should take you to reach your destination. Otherwise, if you were looking for a particular place or thing you could pass it without knowing it.

The table at the bottom of the page can be helpful in estimating how long it takes to walk a mile in four different kinds of situations. Use it to answer the following questions.

a. You are going to walk two miles—one mile through woods and one mile through fields. About how many minutes will you be walking?

_____ minutes

b. You are going to walk along a highway for one mile, then you are going to turn off the highway and walk through woods for one-half mile. About how many minutes will you be walking?

_____ minutes

c. You are going to walk up a mountain for a mile. Then, you are going to walk down the mountain for a mile. Finally, you are going to walk through a field for a mile. About how many minutes will you be walking?

_____ minutes

The Estimated Time It Takes to Walk a Mile Under Certain Conditions	
Place	**Minutes**
Woods (Thin)	30
Highway	20
Fields	25
Forests or Mountains	40

© 2000 by The Center for Applied Research in Education

SECTION

ACTIVITIES

16-1   **Geography Jigsaw Puzzle (Instructor)** 194

16-2   **Crossword: Countries of the World** 195

16-3   **Geography Dioramas (Instructor)** 196

16-4   **Research: Questions and Answers I (Instructor)** 197

16-5   **Research: Questions and Answers II (Instructor)** 198

16-6   **Geography Bingo I (Instructor)** 199

16-7   **Geography Bingo II (Instructor)** 200

16-8   **Geography Bingo III (Instructor)** 201

16-9   **Geography Bingo IV (Instructor)** 202

16-10   **Latitude and Longitude Treasure Hunt** 203

16-11   **"TV Filmstrip" in a Box (Instructor)** 204

16-12   **Unscramble the Names of Countries and Capitals** 205

16-13   **Geography Wordsearch** 206

16-14   **Environmental Riddles** 207

16-15   **Environmental Poetry** 208

GEOGRAPHY JIGSAW PUZZLE

Putting a jigsaw puzzle together can be an interesting and challenging activity for students. And, since the puzzles will be concerned with geographic matters, they can be productive of facts and details. Assembling puzzles can also be conducive to students working together on a common project. Also, creating puzzles requires very little expenditure of money. Geography-centered pictures are easily available. Calendars often have geographic scenes, and most of the pictures on them are attractively colored. Other sources of pictures are old *National Geographic* magazines and brochures from travel agencies.

Suggestions for Conducting the Activity

1.   Obtain pictures from sources such as those mentioned above.
2.   Paste each picture on a piece of cardboard or oak-tag. After the pictures have been attached, trim off any excess cardboard.
3.   Allow the pictures to dry on the cardboard. Then, draw lines on the pictures, and cut the pieces along the lines. An "easy" puzzle may have from 8 to 12 pieces. A moderately difficult puzzle may have 13 or more pieces.
4.   Package the puzzle pieces in a strong envelope. Label the envelope with the title of the picture.

Additional Suggestions

1.   Maps are also suitable as jigsaw puzzles. The map below can serve as an example.
2.   A set of questions for each puzzle may be written. After students have assembled the puzzle they can find the answers to the questions by studying the picture or map. Typical detail questions based on the map below could be, for example, What state has a capital with a girl's name? What state has the name of a president for its capital? Which state extends farthest south?
3.   There is no reason why students can't make their own puzzles either as a class activity or as a homework assignment. In this way a large collection of puzzles can be acquired for use in future classes.

Notes:

1.   Notice that it is not necessary to cut into the small indentations around the perimeter of the map.

2.   A title has been included which is an integral part of the puzzle

3.   The picture below shows how a picture may be partitioned.

CROSSWORD-COUNTRIES OF THE WORLD

This puzzle is different from most crossword puzzles. Each entry has 1–4 letters from its name already printed in the squares of the puzzle. Your challenge is to find the name of one of the countries listed below that fits exactly in the spaces. When you find the name print it on the puzzle.

List of Countries

- CHAD
- ICELAND
- INDIA
- SLOVAKIA
- ALGERIA
- SPAIN
- LIBYA
- NEPAL
- PERU
- ARGENTINA
- JAPAN
- KENYA
- VENEZUELA
- IRAQ
- CHILE
- CUBA
- CHINA
- ISRAEL
- POLAND
- LIBERIA
- IRELAND
- CANADA
- AUSTRALIA
- EGYPT

GEOGRAPHY DIORAMAS

Dioramas can call up a meaningful combination of geography and art skills. Also, they provide opportunities for students to work with their hands and obtain satisfying results. They are simple to make and not very expensive. When completed they are a source of pleasure—something exists that didn't exist before. Unlike pictures, which show things in two dimensions, dioramas are three dimensional. Dioramas allow for the development of research and organizational skills. And, making a diorama can be an opportunity for two students to work together.

Some scenes and topics that lend themselves to geography dioramas are farms and farm workers plowing, harvesting, etc.; mountain scenes with perhaps a cabin in the valley; forest and forest workers; winter sports such as skiing, sledding, and ice-skating; ranch activities such as herding cattle and sheep; water mills and windmills; volcanoes; seashore scenes with lighthouses, piers, ships at sea; roadside fruit and vegetable stands; river scenes (see drawings below of the Nile River and the Congo River); desert scenes. Textbooks will also offer pictures that lend themselves to diorama depictions.

Steps in Making Dioramas

1. A topic is chosen—for example, the depiction of a fishing harbor scene. (See below)

2. A shoe box is obtained.

3. A background is drawn on a piece of strong paper or thin cardboard. The paper should be the size of the back wall of the shoe box and should be neatly glued against the wall. Fluffy cotton clouds, cutout birds, and a yellow sun may be attached to the background.

4. In the foreground, symbols that represent the fishing industry can be made and placed. The symbols may be cardboard cutouts or three dimensional.

The shoe-box diorama shown below shows one way of making a fishing diorama. The two small pictures show river scenes of the Nile and Congo Rivers that also could be made into attractive dioramas.

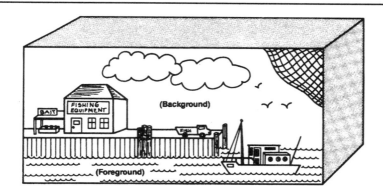

TWO GREAT AFRICAN RIVERS—THE NILE AND THE CONGO

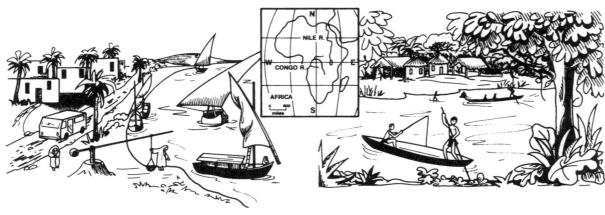

RESEARCH: QUESTIONS AND ANSWERS I

This activity offers an opportunity for students to research in references such as encyclopedias, world almanacs, standard textbooks, and other informational books. The research does not require long essay-type responses; responses rarely require more than a few words.

Ordinarily in research, a question is asked and an answer is to be found. In this activity, however, the answer is given. After studying and researching the answer/statement, students must supply the question that the answer/statement satisfies. An example follows:

STATEMENT

Two smaller rivers come together in North Africa to make this great river, which flows into the Mediterranean Sea.

QUESTION

What is the Nile River?

The bottom of this page and the following page offer sets of statements. Notice that all the statements in a set are on the same topic. Small groups of students could work together on a topic. After each group has researched its statement, the group could ask the class to respond to the statements the group has researched.

One variation of this activity is for students to compose sets of research statements to which other students must respond with appropriate questions. Clarity and briefness should be emphasized in the composition of the statements. This factor brings a language-arts element into the activity.

SETS OF QUESTIONS

Rivers of the World
Statements

1. The capital of France is located on this river.

2. This, the greatest river in South America, flows eastward toward the Atlantic Ocean.

3. Two rivers, the Monongahela and the Allegheny, join together to form this great river, which flows to the Mississippi River.

4. This river forms part of the boundary between the United States and Mexico.

5. Close to St. Louis, Missouri, this great river from the northwest joins the Mississippi River.

6. London, the capital of the United Kingdom of Great Britain and Northern Ireland, is located on this river.

7. This, the greatest river in Italy, runs in the north of Italy and flows into the Adriatic Sea.

8. Three of Europe's capital cities—Belgrade, Budapest, and Vienna—are located on this river, which flows eastward into the Black Sea.

Questions

1. What is the River Seine?

2. What is the Amazon River?

3. What is the Ohio River?

4. What is the Rio Grande?

5. What is the Missouri River?

6. What is the Thames River?

7. What is the Po River?

8. What is the Danube River?

RESEARCH: QUESTIONS AND ANSWERS II

Waters of the World
Statements

1. There are five Great Lakes, but only this one is entirely within the United States.

2. This large bay in Europe is north of Spain and west of France.

3. Sicily, a large island off the "toe" of Italy, is separated from the mainland by this narrow strait.

4. This beautiful South American lake is about two miles above sea level and lies on the border between Peru and Bolivia.

5. This great bay lies north of Canada and makes a great indentation in Canada's northern coast.

6. This sea in Europe is bounded by Italy, Corsica, Sardinia, and Sicily.

7. A favorite vacation spot for many Americans, this long lake forms part of the border between Vermont and New York.

8. The peninsula on which Sweden is located is long and narrow; this body of water separates the country from Finland.

Questions

1. What is Lake Michigan?

2. What is the Bay of Biscay?

3. What is the Strait of Messina?

4. What is Lake Titicaca?

5. What is Hudson Bay?

6. What is the Tyrrhenian Sea?

7. What is Lake Champlain?

8. What is the Gulf of Bothnia?

Mountains of the World
Statements

1. This range of mountains along the west coast of South America has been called "The Backbone of South America."

2. These mountains which run from Georgia to Maine, were great obstacles to westward-moving Americans during colonial times.

3. This range of mountains separating Norway and Sweden serves as a natural boundary between the two countries.

4. This mountain, highest in the world, is part of the Himalaya Mountains in Asia.

5. These mountains are located between the Black Sea and the Caspian Sea.

6. These mountains in northwest Africa at the western end of the Sahara Desert were named after a Greek god.

Questions

1. What are the Andes Mountains?

2. What are the Appalachian Mountains?

3. What are the Kjolen mountains?

4. What is Mount Everest?

5. What are the Caucasus Mountains?

6. What are the Atlas Mountains?

GEOGRAPHY BINGO I

Learnings that are not repeated/rehearsed will become dim in learners' minds; they need to be reinforced frequently. Among those things most easily forgotten are facts. Facts are important; they are the building blocks of thinking. As Sir Joshua Reynolds once wrote,

<u>Thinking Point</u>
Invention, strictly speaking,
is little more than a new combination
of those images which have been
previously gathered and deposited
in the memory. Nothing can be made
of nothing. They who have laid
up no materials can produce no
combinations.

So, there is a place in today's classrooms for drill and rehearsal. Repetition does not necessarily result in boredom, especially if the presentation of the repeated material varies. The Geography Bingo game described on these pages is one such exciting and challenging way to rehearse facts related to geography.

The card shown below is an example of how students' cards may be arranged before the start of the Bingo game.

SAMPLE CARD

Greenland	Panama Canal	China	Lake Michigan	Cape of Good Hope
Gulf of Mexico	Egypt	Mt. McKinley	Antarctica	Rhode Island
Saudi Arabia	Tunisia	Appalachian Mountains	Bering Strait	Persian Gulf
Red Sea	Niagara Falls	Santiago	Rio Grande	Himalaya Mountains
Cuba	Ural Mountains	Ottawa	Florida	Caspian Sea

GEOGRAPHY BINGO II

Suggestions for Playing Geography Bingo

Procedure One

Play the game the same way that Bingo is played by most people. However, following are adaptations to make for this version of Bingo.

1. Photocopy the game grid on the facing page and distribute.

2. Direct student attention to the word list below the grid. In each square of the grid, they are to carefully print a different word or phrase from the list. The words should be selected and placed randomly. In this way, no two grids in the class will be the same.

3. Distribute tokens (paper clips, small cardboard squares, etc.) to be used to cover the squares.

4. As you randomly read words from the list, students should cover the squares on their grids that contain those words.

5. When all five squares in any row—horizontal, vertical, or diagonal—are covered on a student's grid, he/she should raise a hand and call out, "Bingo!"

6. To play succeeding games have students clear their grids of tokens.

Note: Each of the terms in the list can be copied on half of a 3" × 5" index card. Before each game, the cards can be shuffled and then read by the teacher or game "caller." This will ensure that a variety of terms are called, and it will facilitate checking the accuracy of the winner's grid.

Procedure Two

This procedure is more difficult for students; however, it can also be more profitable in terms of geographic understandings and retention.

1. Photocopy and distribute page 202. Students randomly choose items (answers) that follow the statements. Then, they write one item in each square on their grid.

2. The instructor reads one of the descriptions on page 202, but without the final identification.

3. Without referring to their copies of page 202, students recall the appropriate identification. If they have the answer on their grid, they cover it.

4. The first student who has a five-square row completed—horizontally, vertically, diagonally—is the winner. Carefully review the descriptions and identifications with the class.

Note 1: Before playing this version of the game, teach the 48 listings. Go over the listings several times, and point out places on a wall map. Then, play the game. After the game is over, go over the listings again. In time, students will become very good at making the proper relationships.

Note 2: Instructors can make their own place/identification listings that may be more closely related to what has been or is being studied.

Procedure Three

Make the listings on page 202 a research problem. That is, photocopy the listings without the answers. Then, perhaps once a week, have students do research to find the places that fit the descriptions. Again, as suggested above, it may be more advantageous to make your own customized listings that are in agreement with what is being currently studied. This kind of researching task will acquaint students with various reference books and maps.

An alternative approach would be to furnish students with the place-names and then have them furnish the definitions. Again, the students will need to refer to reference texts and maps. In addition, they will utilize language-art skills in writing the descriptions.

© 2000 by The Center for Applied Research in Education

GEOGRAPHY BINGO III

PLACE-NAME LISTINGS

Sweden	Mt. Aconcagua	Canada	China
Brazil	Appalachian Mountains	Andes Mountains	Japan
Atacama Desert	Red Sea	Iceland	Gulf Stream
Mt. Kilimanjaro	Madagascar	Hawaii	Nile River
Niagara Falls	Santiago	Panama Canal	France
Saudi Arabia	Sacramento	Egypt	Norway
Texas	Caspian Sea	Arctic Ocean	Strait of Gibraltar
Alaska	Florida	Australia	Gulf of Mexico
Mt. Everest	Ural Mountains	Greenland	Strait of Magellan
Bering Strait	Italy	Lake Michigan	Russia
Rio Grande	Great Britain	Mississippi River	Sahara Desert
Amazon River	Cuba	Cape of Good Hope	Argentina

GEOGRAPHY BINGO IV

The questions that follow are numbered for easy referral; however, they can be utilized in any order.

1. Northern neighbor of the United States: Canada
2. South America's largest mountain range: Andes Mountains
3. Island country in the North Atlantic Ocean: Iceland
4. United States state in the Pacific Ocean: Hawaii
5. Second largest country in South America: Argentina
6. World's most northern ocean: Arctic Ocean
7. A continent with only one country: Australia
8. A very large island in the North Atlantic Ocean: Greenland
9. The only Great Lake entirely within the United States: Lake Michigan
10. The largest river in the United States: Mississippi River
11. The largest country entirely within the continent of Asia: China
12. An island nation off the east coast of Asia: Japan
13. Large body of water south of the United States: Gulf of Mexico
14. Current of warm water that flows northeast across the Atlantic: Gulf Stream
15. A river that begins in the mountains of Africa and flows north into the Mediterranean Sea: Nile River
16. Water passage in South America with explorer's name: Strait of Magellan
17. A very large country that occupies part of Asia and Europe: Russia
18. A country that occupies the eastern part of the Scandinavian Peninsula: Sweden
19. This country's capital is Oslo: Norway
20. Spain is on this country's south; the Bay of Biscay is on its west: France
21. The world's largest desert: Sahara Desert
22. Country in which the Suez Canal is located: Egypt
23. A canal that connects the Atlantic and Pacific Oceans: Panama Canal
24. Narrow passage at the western end of the Mediterranean Sea that connects it with the Atlantic Ocean: Strait of Gibraltar
25. A cape at the southern end of Africa: Cape of Good Hope
26. Largest country in South America: Brazil
27. South American desert located between Peru and Chile: Atacama Desert
28. Highest mountain in Africa: Mt. Kilamanjaro
29. A great waterfall partly in New York and partly in Canada: Niagara Falls
30. A state with Oklahoma on its north and New Mexico on its west: Texas
31. The United States most northern state: Alaska
32. The world's highest mountain: Mt. Everest
33. Water passage that connects the Arctic Ocean and the Bering Sea: Bering Strait
35. South America's largest river: Amazon River
36. South America's highest mountain: Mt. Aconcagua
37. Mountain range that stretches from Georgia to Maine: Appalachian Mountains
38. Sea that separates Egypt from the Arabian Peninsula: Red Sea
39. Large island off the southeast coast of Africa: Madagascar
40. The capital of the South American country of Chile: Santiago
41. The capital of California: Sacramento
42. The country that occupies most of the Arabian Peninsula: Saudi Arabia
43. The country about 90 miles south of the southern tip of Florida: Cuba
44. The island on which England, Scotland, and Wales are located: Great Britain
45. The country whose shape has the appearance of a boot and which extends into the Mediterranean Sea: Italy
46. The mountains that separate Europe and Asia: Ural Mountains
47. State with coastlines on the Atlantic Ocean and Gulf of Mexico: Florida
48. The worlds' largest inland sea: Caspian Sea

LATITUDE AND LONGITUDE TREASURE HUNT

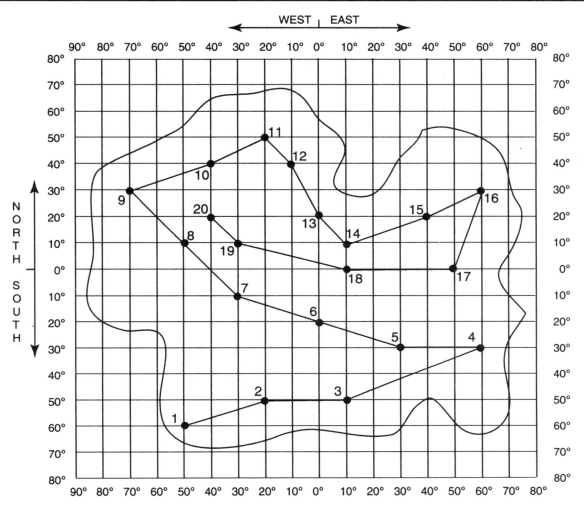

If you follow the line on the map from number 1 to number 20 you will come to the spot where a great treasure is buried. But before you can have the treasure you must be able to identify the latitude and longitude of all the numbered places on the map.

Directions

1. Start at 60°S and 50°W.

2. Follow the line shown by dots and numbers.

3. At each numbered dot determine the latitude and longitude. Then write the location beside the same number in the next column.

4. Number 1 has been recorded to help you get started.

5. When you reach the end of your treasure hunt, feel good! Your treasure is that you developed greater skill in matters of latitude and longitude.

Latitude and Longitude

1. 60°S–50°W _____   11. _____
2. _____   12. _____
3. _____   13. _____
4. _____   14. _____
5. _____   15. _____
6. _____   16. _____
7. _____   17. _____
8. _____   18. _____
9. _____   19. _____
10. _____   20. _____

"TV FILMSTRIP" IN A BOX

A student-made filmstrip is basically a series of frames depicting a particular topic. It is an activity that is a productive combination of mental and physical (hands-on) effort. This activity is potentially conducive to teaching and learning geography because it combines visual images with text, so students can see the geographical elements of various topics.

Students derive satisfaction from the finished product. Numerous skills are employed, and significant subject matter understandings are achieved.

Following are some suggestions relative to implementing a filmstrip activity.

Frames for the Filmstrip

1. Use 8½" × 11" paper.

2. Limit the number of frames to be drawn to 12.

3. The frames should be on a particular theme. Following are some typical themes.

a. <u>Transportation</u>: Frames could include types of vehicles; transportation accessories such as bridges, tunnels, and roads; traffic signs—stop signs, speed limit signs, and signs about distance and direction.

b. <u>Countries</u>: Frames could include maps showing political and physical components, unusual national elements such as windmills and dikes in the Netherlands, deserts and volcanoes within countries.

c. <u>Industries</u>: Frames could depict facets of industrial processes—lumbering, mining, fishing, farming, and manufacturing.

d. <u>Physical and Human-Made Features</u>: Frames could show rivers, mountains, and national parks.

e. <u>History with Emphasis on Geography</u>: Information is readily available for frames about the geographical elements of the American Revolution, the Civil War, World Wars I and II, the westward movement, life on the Great Plains, etc.

f. <u>Recreation</u>: Frames can depict the seashore; camping; boating; sports such as skiing, ice-skating, mountain-climbing, and hiking.

Constructing the Filmstrip Box

1. The filmstrip box, shown in the next column, can be made by students. The basic materials needed to construct the box are easily obtainable and inexpensive. What is needed?

• A sturdy cardboard box about 12" high, wide, and long

• Two wooden rollers, or dowels, for mounting the filmstrip (The rollers should be long enough to go through the box and still project ends for grasping and turning. Cut broom or mop handles make good dowels. Also, dowels about ¾" in diameter are obtainable in hardware stores or lumber stores.)

• White paper about 8½" × 11"

• Scotch tape

2. Cut a hole in the box that is about 1" smaller than the dimensions of the paper. In this way the margins of the frames on the strip will not be seen when the filmstrip is displayed.

3. After the frames of the filmstrip have been drawn, scotch-tape them together in sequence.

4. Punch holes in the sides of the box so that the rollers can slip through. Keep the holes smaller than the rollers so that they will be held tight.

Additional Suggestions

1. Pictures, cartoons, graphs, tables, etc., can be cut out of newspapers and magazines and then attached to the paper.

2. Captions and brief narratives can be part of a frame.

3. Narratives for the filmstrips can be written and recorded on audio tape.

4. Students can work in groups of two or three. Several different filmstrips can be prepared within one classroom.

Name: _____ Date: _____

UNSCRAMBLE THE NAMES OF COUNTRIES AND CAPITALS

Scrambled Countries

1. N I A D I _____

2. K M N R A E D _____

3. R H Y G A U N _____

4. L C H E I _____

5. Z A R B L I _____

6. N E E S D W _____

7. N R F A E C _____

8. U A T S A I R _____

9. G U I L B E M _____

10. A A A C N D _____

11. A A A B L N I _____

12. A S N P I _____

13. U E P R _____

14. K R Y E T U _____

15. O O L M C A I B _____

16. P A I O H E T I _____

17. T V M E N A I _____

18. L N F D A I N _____

19. N A S U D _____

20. L I B O I A V _____

Note: The countries in alphabetical order are ALBANIA, AUSTRIA, BELGIUM, BOLIVIA, BRAZIL, CANADA, CHILE, COLOMBIA, DENMARK, ETHIOPIA, FINLAND, FRANCE, HUNGARY, INDIA, PERU, SPAIN, SUDAN, SWEDEN, TURKEY, VIETNAM

Scrambled Capitals of the United States

1. N E T R O T N _____

2. B N Y A A L _____

3. T N E O A R M A C S _____

4. T S O B N O _____

5. M A S I B C K R _____

6. K E A P T O _____

7. G A A U U T S _____

8. V E L I L H N S A _____

9. S S A A A L L T H E E _____

10. G I E A L R H _____

11. O O L U U N H L _____

12. N A E L H E _____

13. L D E F I G S P R N I _____

14. N E Y E N C E H _____

15. A A A T T N L _____

16. V I D P N E E C R O _____

Note: The capitals in alphabetical order are: ALBANY, ATLANTA, AUGUSTA, BISMARCK, BOSTON, CHEYENNE, HELENA, HONOLULU, NASHVILLE, PROVIDENCE, RALEIGH, SACRAMENTO, SPRINGFIELD, TALLAHASSEE, TOPEKA, TRENTON

Name: _____ **Date:** _____

GEOGRAPHY WORDSEARCH

Here is an opportunity for you to make a GEOGRAPHY WORDSEARCH puzzle. The directions for making it are simple.

1. Make a list of about 25 words related to geography. Try to have from 10 to 14 words that will run across the puzzle and 10 to 12 words that will run down the puzzle.
2. Following are topics related to geography. Within the topics you will be able to think of related words.
   - Political and geographical entities such as continents, countries, and cities
   - Land and water forms such as mountains, deserts, rivers, volcanoes, lakes, oceans, bays, gulfs
   - Weather elements such as rain, hurricanes, hail
   - Agricultural products such as oranges, coconuts, wheat
   - Human-made structures such as dams, bridges, famous buildings
   - Natural landmarks such as geysers, waterfalls, canyons
3. Fit the words into the spaces of your puzzle. Be sure to have some of your words intersect.
4. After you have placed all the words in your list, fill in the leftover spaces with random letters.
5. Alphabetically list the hidden words at the bottom of the puzzle.
6. Give your puzzle to a classmate to solve.

ALPHABETICAL LISTING OF WORDS

_____ _____ _____ _____ _____

_____ _____ _____ _____ _____

_____ _____ _____ _____ _____

_____ _____ _____ _____ _____

_____ _____ _____ _____ _____

Name: _____ Date: _____

ENVIRONMENTAL RIDDLES

Environmental matters are serious; however, a few environmental riddles and unfinished rhymes can help lighten the problems and issues facing us.

1. Try to find humorous answers to the riddles below. Your instructor will have some suggested solutions, but that doesn't mean that yours are wrong. Yours may be even better!

• What kind of flowers were fought by the armored knights of old? _____

_____

• What kind of insects did the knights fight?

_____

• What kind of flowers do female lions like best?

_____

• What kind of flowers do tigers like best?

_____

• What did the elephant say to the tourist traveling through Africa? _____

_____

• What did the father buffalo say to his son who was going on a trip? _____

_____

• What do rattlesnakes get for their very young children? _____

_____

• What did the old, worn-out tire say as it was going through the recycling plant? _____

_____

• How are a mature male deer and a dollar bill alike? _____

_____

• What did the glass jar say as it was going through the recycling plant? _____

_____

• What did the flower say to the bee? _____

_____

• What did the rechargeable battery say to the young boy? _____

_____

2. Try to think of one or two riddles that are in some way connected to the environment. Think of various topics that you have learned about. Your riddle may be a play on words, a "far-out" comparison, and so on. Some topics: litter, smog, air, soil, animals, acid rain, forests, birds, and so on.

Riddle: _____

_____

Answer: _____

Riddle: _____

_____

Answer: _____

3. Here is an incomplete poem; the last line is missing. Try to complete the poem so that the last line rhymes with the line above it.

Air Pollution

What has air pollution done?
Can't see the stars; can't see the sun,
Pits our statues; ruins our lungs!
Can even taste it on our tongues.
The pollution-solution as you can see,

_____

ENVIRONMENTAL POETRY

Haiku Poetry

Haiku poetry originated in Japan. It has a very simple three-line construction, but it can express meaningful thoughts. Because each line has a limited number of syllables, the writer must choose each word carefully. The result is usually a better poem.

Many haikus describe things found in nature, so haiku is especially suitable for environmental issues. Here are two examples.

The Earth

Earth—so very old, (5)
Flowers, trees, rocks, sky, water, (7)
Strong—yet so fragile. (5)

Redwood Tree

Tall, strong, but tender,
On Earth a thousand seasons;
Blunt the axe that strikes.

Note: The numeral in parentheses after each line tells the number of syllables. Count them.

Although haikus are generally about the seasons and nature, other environmental matters—energy, pollution, animals, mountains, deserts, oceans, etc.—can serve as topics.

In the two spaces below, write your own haikus. A helpful procedure would be to first write your poems on scrap paper; then, when you have them written to your satisfaction, transfer them to the lines provided below. Remember: The first line has 5 syllables; the second line has 7 syllables; the third line has 5 syllables.

Title: _____

_____

_____

_____

Title: _____

_____

_____

_____

Rhyming Poem or Free Verse (non-rhyming)

At the bottom of this column, write a poem in any form that you find suitable. Your poem can rhyme, but it doesn't have to. It should have a title and at least four lines. It can be very serious, or it can be light in nature as is the poem printed below. Finally, your poem should have something to do with the environment or geography.

Little Fish

One, two, three, four, five,
Once I caught a fish alive.
Six, seven, eight, nine, ten,
Then, I let him go again.
"Why did you let him go?"
"'Cause he needed time to grow!"

Here are some suggestions for topics:
• the sea
• endangered animals
• trees, forests
• polluted skies
• polluted water
• growing population
• soil erosion
• recycling (for example, its importance to saving forests, minerals, etc.)

Title: _____

_____

_____

_____

_____

_____

SECTION 17

OUTLINE MAPS

17–1　North America (labeled) 210

17–2　North America (unlabeled) 211

17–3　USA (labeled) 212

17–4　USA (unlabeled) 213

17–5　Middle America (labeled) 214

17–6　Middle America (unlabeled) 215

17–7　South America (labeled) 216

17–8　South America (unlabeled) 217

17–9　South Asia (labeled) 218

17–10　South Asia (unlabeled) 219

17–11　Nations of the Former USSR (labeled) 220

17–12　Nations of the Former USSR (unlabeled) 221

17–13　Australia and New Zealand (labeled) 222

17–14　Australia and New Zealand (unlabeled) 223

17–15　Africa (labeled) 224

17–16　Africa (unlabeled) 225

17–17　Europe (labeled) 226

17–18　Europe (unlabeled) 227

© 2000 by The Center for Applied Research in Education

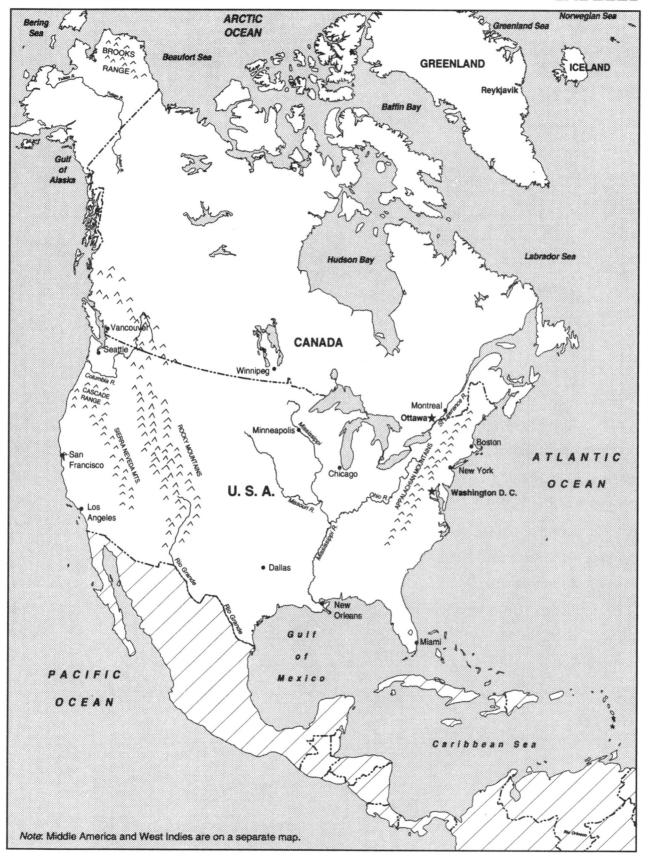

Note: Middle America and West Indies are on a separate map.

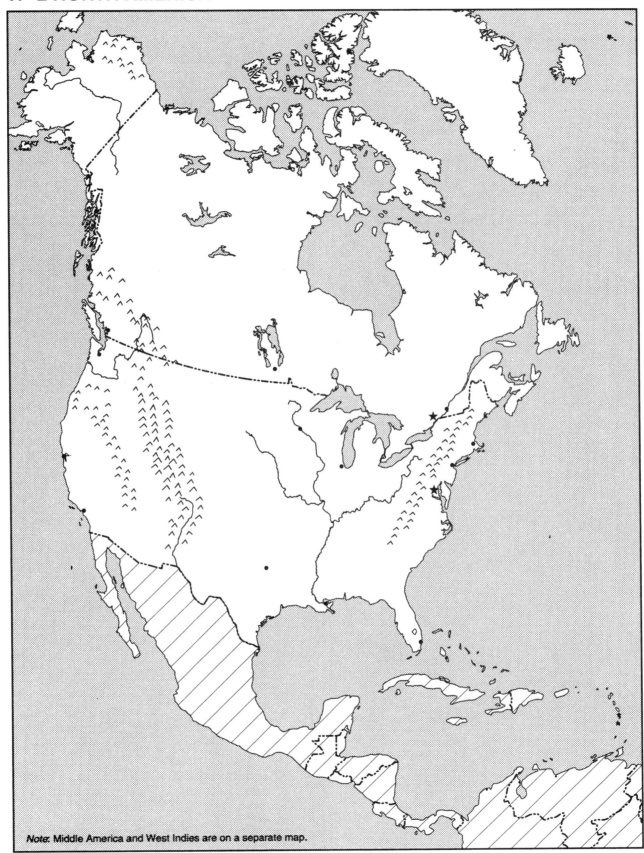

Note: Middle America and West Indies are on a separate map.

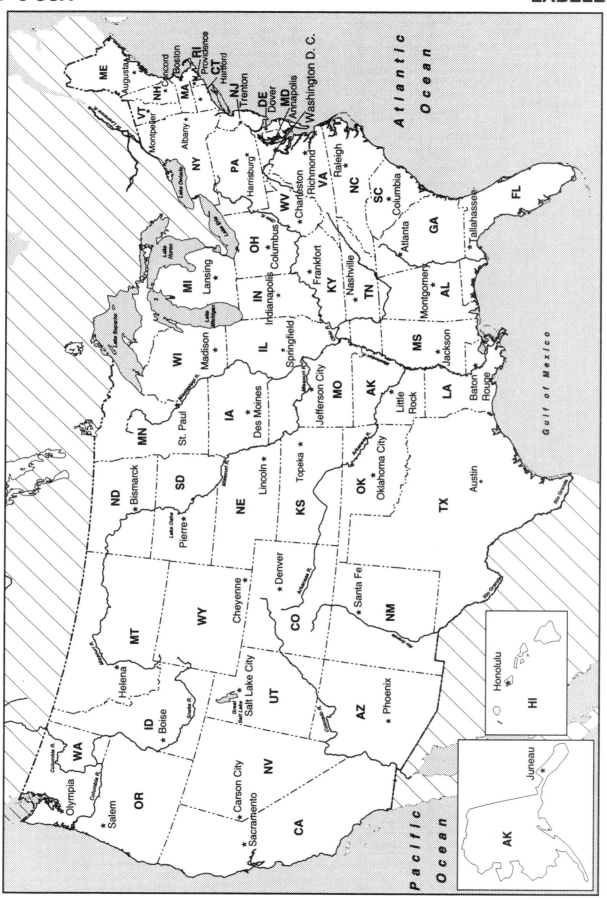

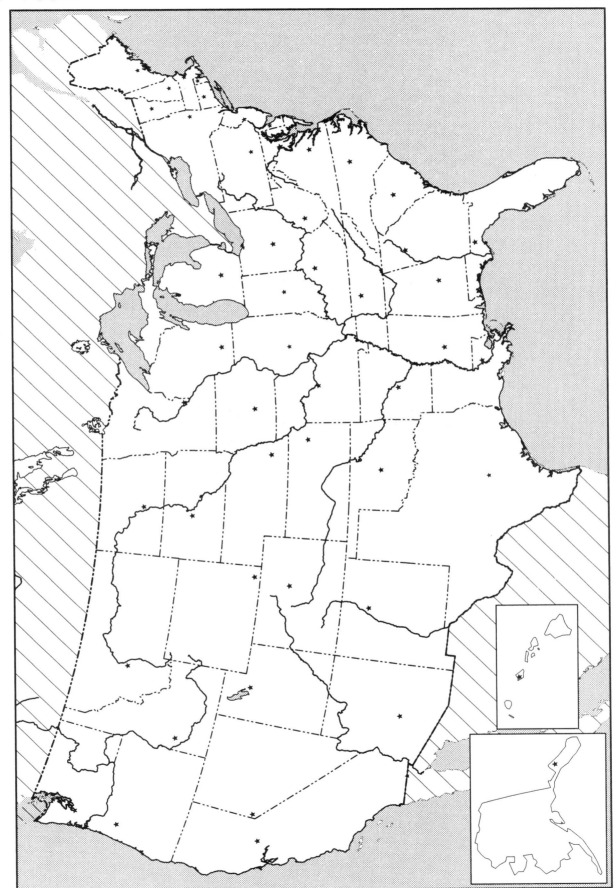

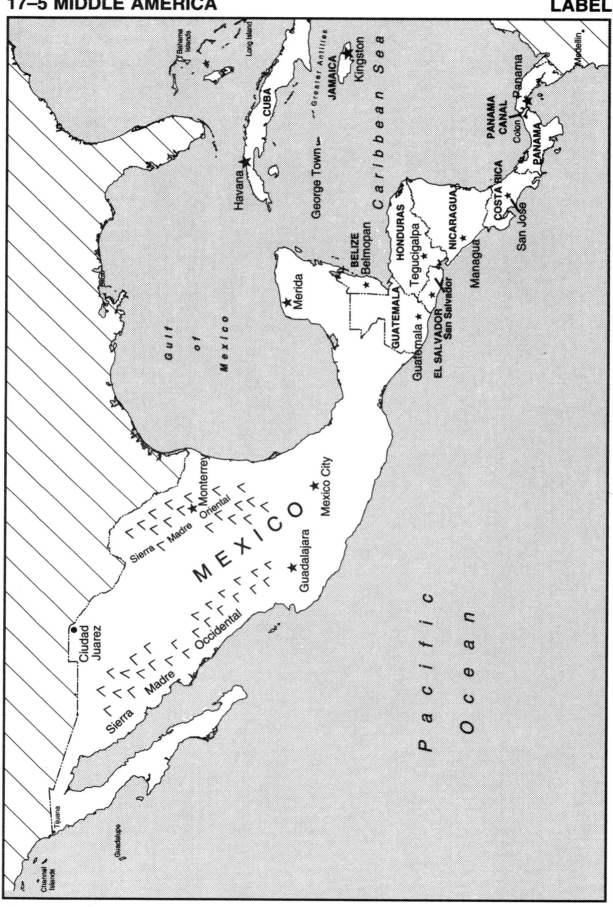

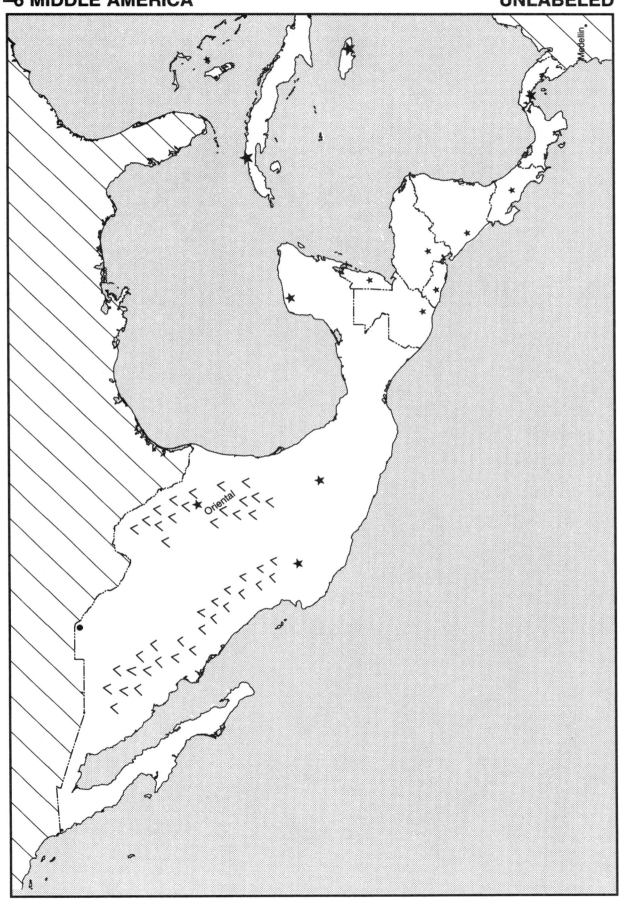

Medellin.

Oriental

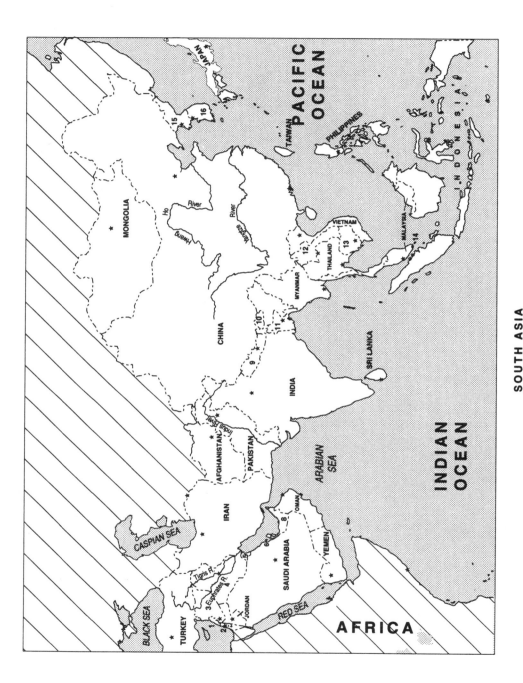

SOUTH ASIA

AFGHANISTAN
(Kabul)
BAHREIN: 6
(Manama)
BANGLADESH: 11
(Dhaka)
BHUTAN: 10
(Thimphu)
CHINA
(Bejiing)

INDIA
(New Delhi)
INDONESIA
(Jakarta)
IRAN
(teheran)
IRAQ: 4
(Baghdad)
ISRAEL: 2
(Jerusalem)

JAPAN
(Tokyo)
JORDAN
(Amman)
KAMPUCHEA: 13
(PhnomPenh)
KUWAIT: 5
(Kuwait)
LAOS: 12
(Vientiane)

LEBANON
(Beirut)
MALASIA
(Kualo Lumpur)
MONGOLIA
(Ulan Bator)
MYAMAR
(Yangon)
NEPAL: 9
(Kathmandu)

NORTH KOREA: 15
(Pyongyang)
OMAN
(Muscat)
PAKISTAN
(Islamabad)
PHILIPPINES
(Manila)
QATAR: 7
(Doha)

SAUDI ARABIA
(Riyadh)
SINGAPORE: 14
(Singapore)
SOUTH KOREA: 16
(Seoul)
sri lanka
(Colombo)
SYRIA: 3
(Damascus)

TAIWAN
(Taipei)
TURKEY
(Ankara)
UNITED ARAB
EMIRATES: 8
(Abu Dhabi)
VIETNAM
(Hanoi)
YEMEN
(Sana)

© 2000 by The Center for Applied Research in Education

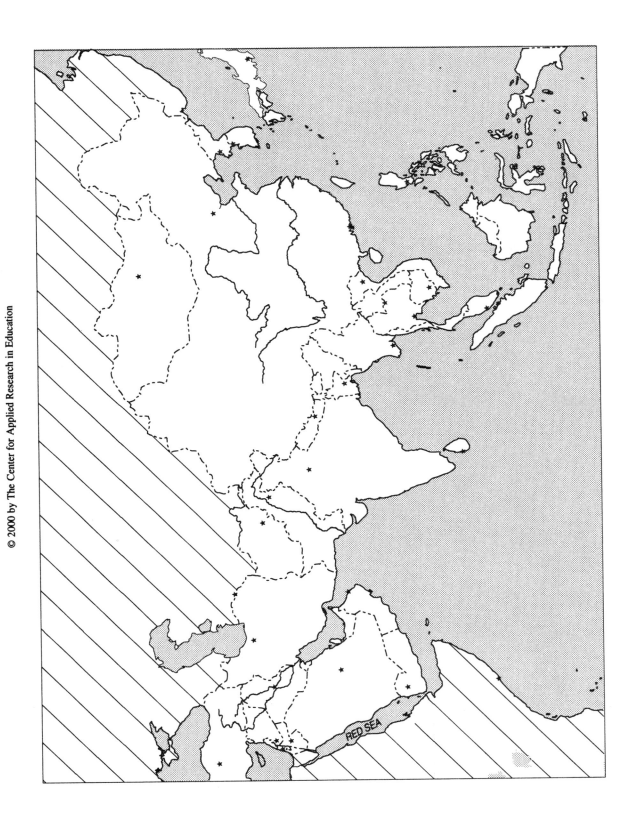

RED SEA

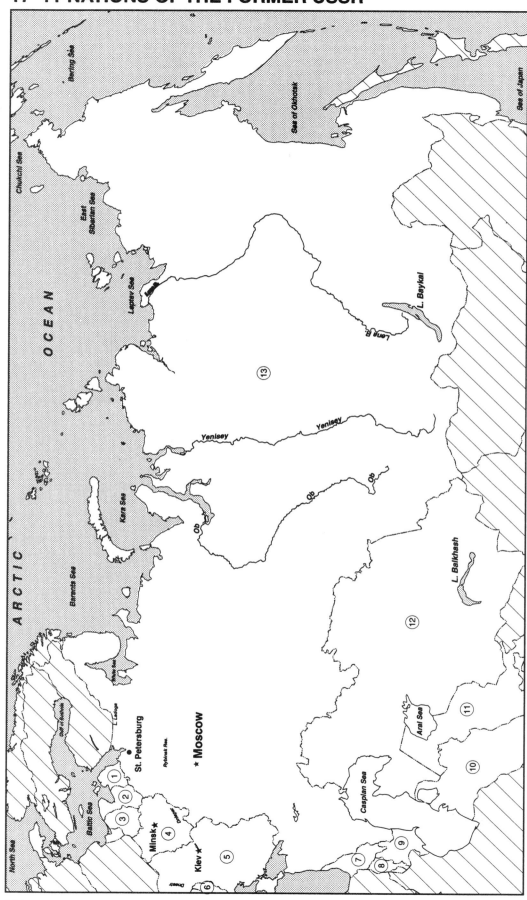

1. ESTONIA
2. LATVIA
3. LITHUANIA
4. BELARUS
5. UKRAINE

6. MOLDAVO
7. GEORGIA
8. ARMENIA
9. AZERBAIJAN

10. TURKMENISTAN
11. UZBEKISTAN
12. KAZAKSTAN
13. RUSSIA

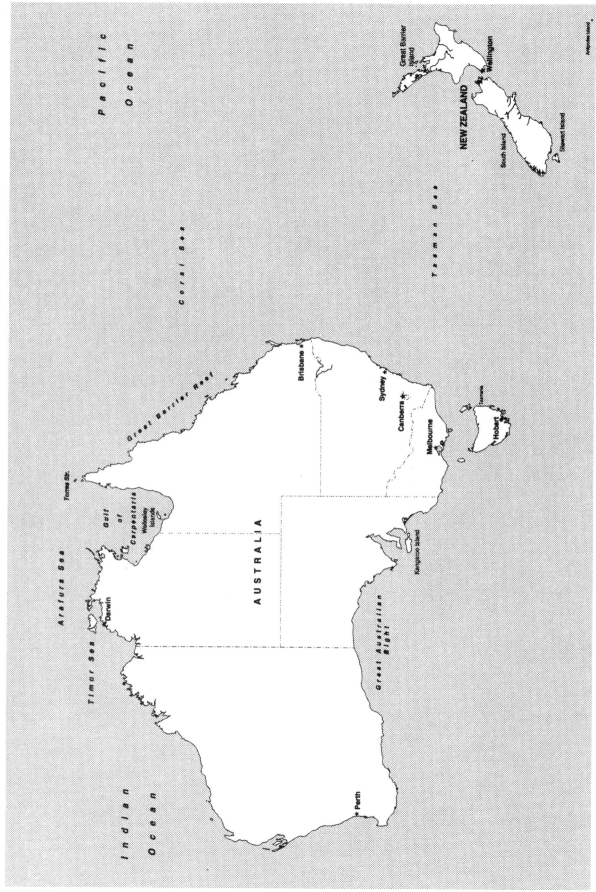

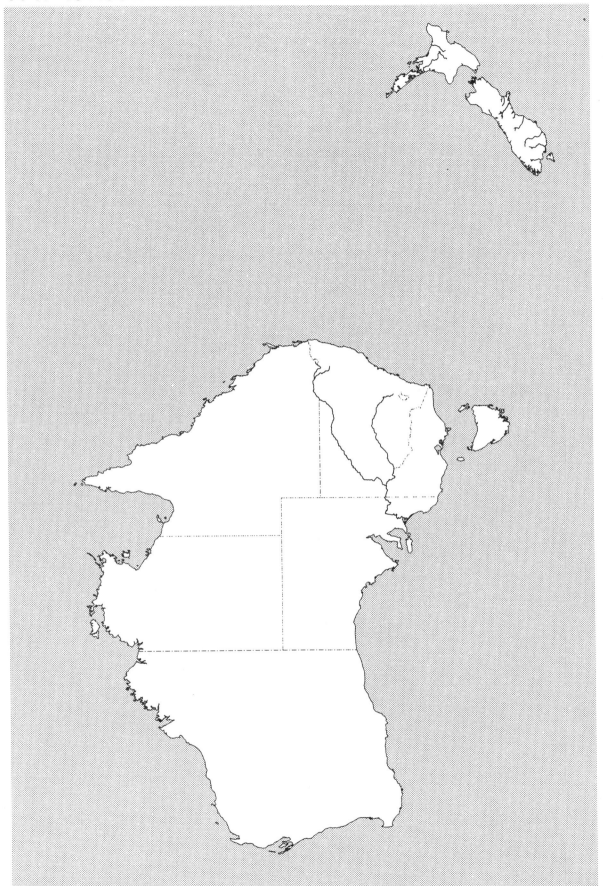

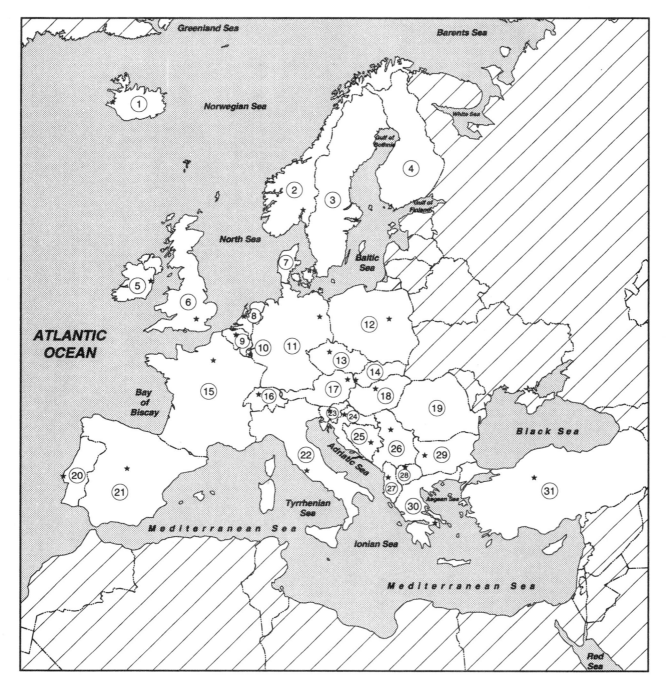

1. ICELAND (Reykjavik)
2. NORWAY (Oslo)
3. SWEDEN (Stockholm)
4. FINLAND (Helsinki)
5. IRELAND (Dublin)
6. U. KINGDOM (London)
7. DENMARK (Copenhagen)
8. NETHERLANDS (Amsterdam)
9. BELGIUM (Brussels)

10. LUXEMBURG (Luxemburg)
11. GERMANY (Berlin)
12. POLAND (Warsaw)
13. CZECH (Prague)
14. SLOVAKIA (Bratislava)
15. FRANCE (Paris)
16. SWITZERLAND (Bern)
17. AUSTRIA (Vienna)
18. HUNGARY (Budapest)

19. ROMANIA (Bucharest)
20. PORTUGAL (Lisbon)
21. SPAIN (Madrid)
22. ITALY (Rome)
23. SLOVENIA (Ljubljana)
24. CROATIA (Zagreb)
25. BOSNIA (Sarajevo)
26. YUGOSLAVIA (Belgrade)
27. ALBANIA (Tirane)

28. MACEDONIA (Skopje)
29. BULGARIA (Sofia)
30. GREECE (Athens)
31. TURKEY (Ankara)

Note: Members of the former
        USSR not included

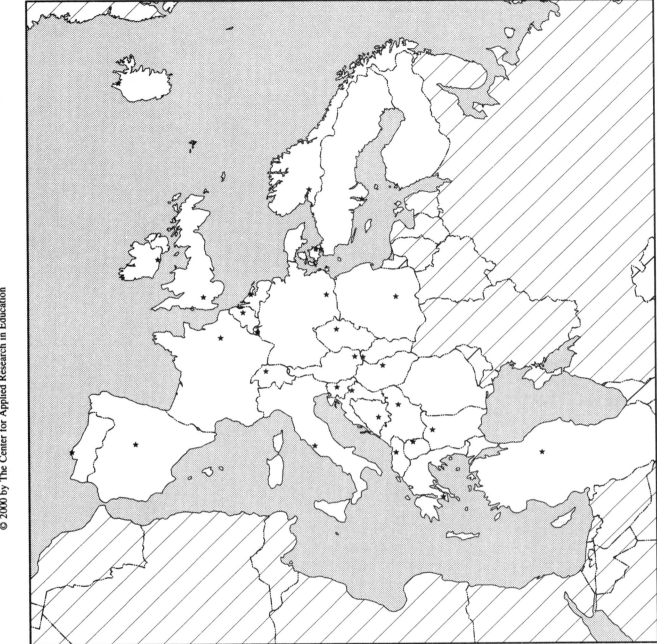

PACIFIC

OCEAN

JAPAN

Bering Sea

Sea of Japan

Gulf of Alaska

Yellow
Sea

South
China
Sea

MEXICO

**NORTH
AMERICA**

CANADA

ARCTIC

MONGOLIA

RUSSIA

CHINA

U. S. A.

OCEAN

ASIA

Hudson
Bay

INDIA

GREENLAND

Aral Sea

Caspian Sea

Norwegian
Sea

ATLANTIC

Baltic
Sea

North
Sea

EUROPE

Black Sea

SAUDI ARABIA

OCEAN

Red Sea

Mediterranean Sea

AFRICA

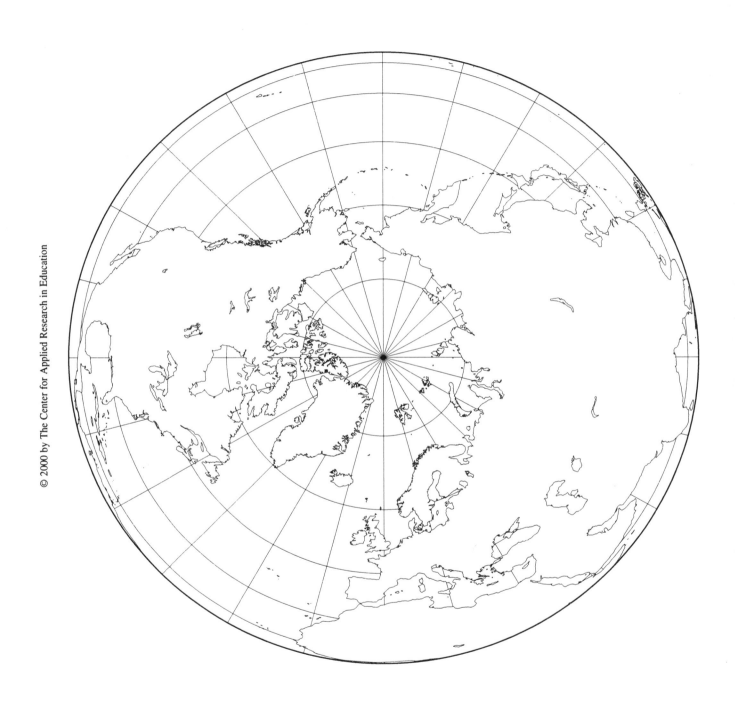

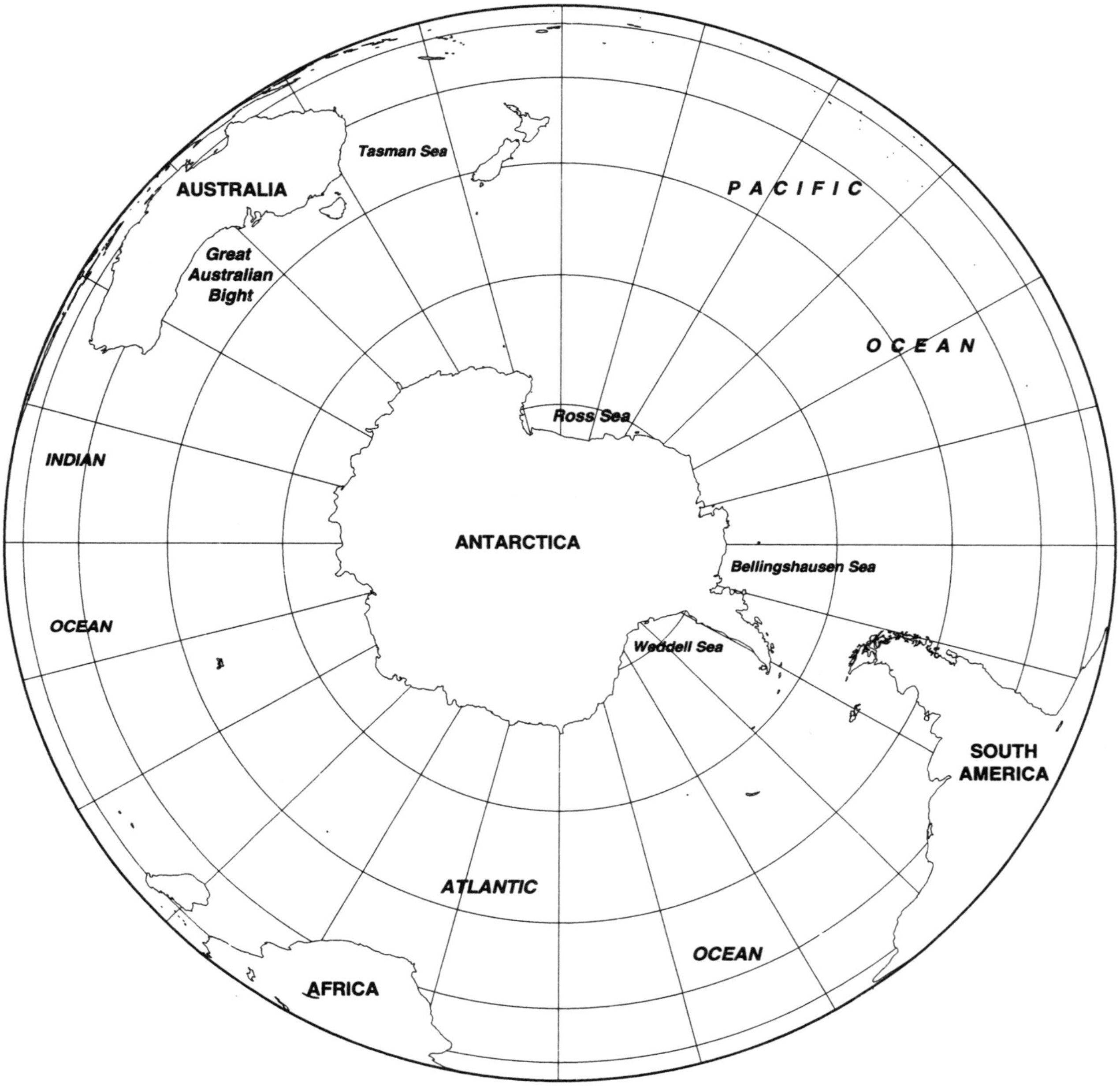

AUSTRALIA

Tasman Sea

PACIFIC

Great
Australian
Bight

OCEAN

INDIAN

Ross Sea

ANTARCTICA

Bellingshausen Sea

OCEAN

Weddell Sea

SOUTH
AMERICA

ATLANTIC

OCEAN

AFRICA

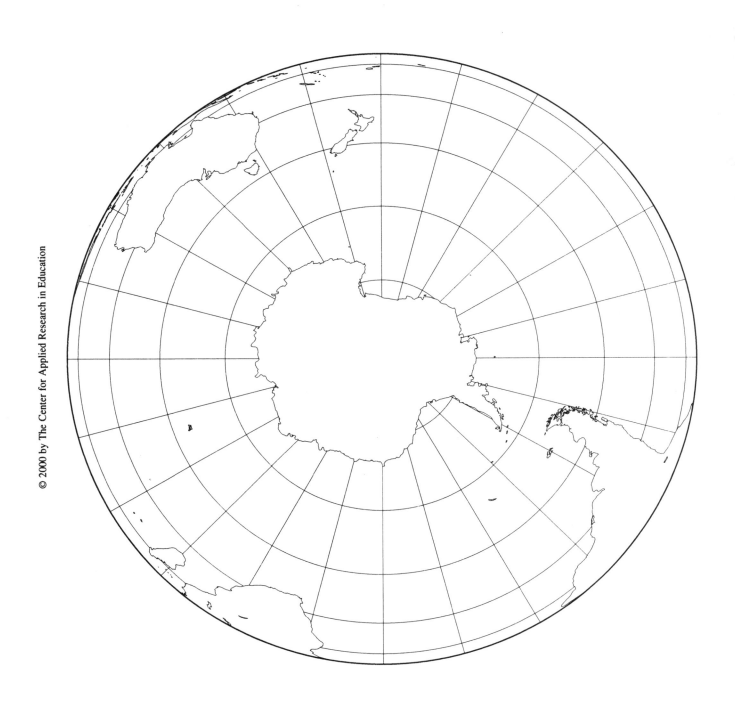

ANSWER KEY

SECTION 1: BASIC MAP READING SKILLS

1-2   DRAWING A MAP

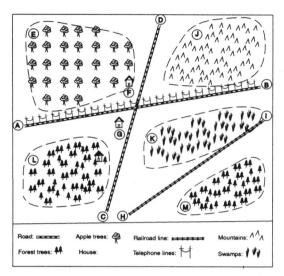

1-4   A MAP WITH NATURAL AND HUMAN-MADE FEATURES

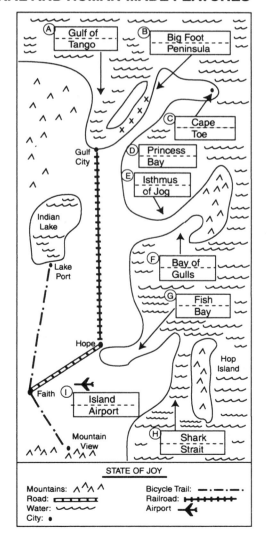

1-6 IDENTIFYING LAND, WATER, AND SKY FEATURES

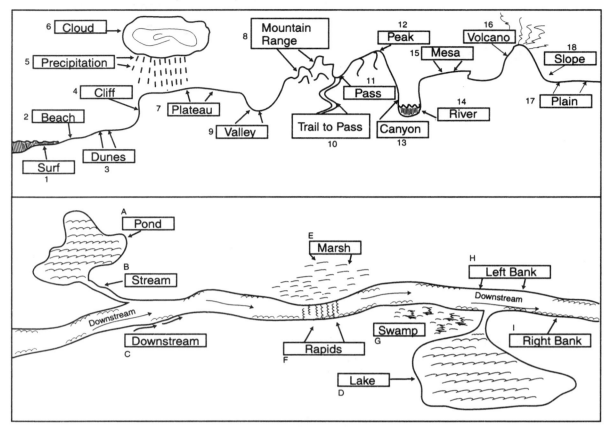

1-8 MAP SYMBOLS FOR THINGS THAT CANNOT BE SEEN

1. ABRA and CADABRA
5. Students will choose their own names.

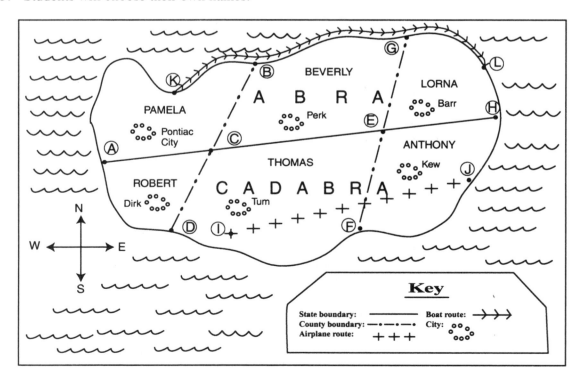

1-10 USING DIRECTIONS TO LOCATE CITIES

CITY	DIRECTION FROM THE CENTER OF THE CIRCLE							
	N	S	E	W	NE	SE	SW	NW
Preston								√
Joplin								√
Bolton				√				
Lahaska		√						
Carlton			√					
Far Hills						√		
Pineville						√		
Lyons		√						
Delta					√			
Morton							√	
Newton	√							
Holt	√							
Woodton						√		
Alpha							√	
Fisk					√			

1-11 COMPLETING A DIRECTION MAZE

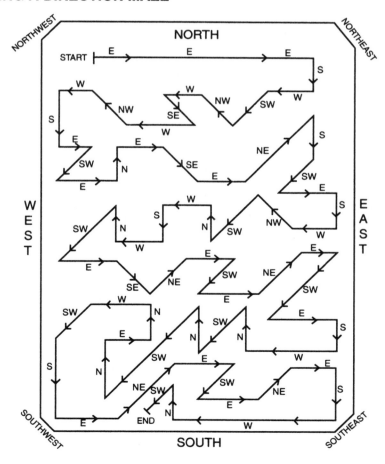

235

1-12 FINDING YOUR WAY AROUND A TOWN

1.a. West
   b. North
2.a. West
   b. East
3. Swim Pool
4. High School
5. Drug Store

6. He was driving the wrong way on a one-way street.
7.a. Walk south one block.
   b. Turn east and walk to the second establishment, the restaurant.
8.a. Walk south two blocks.
   b. Turn west and walk to the first establishment in the third block, the theatre.

1-14 USING A SCALE OF MILES TO FIND DISTANCES ON MAPS

1.a. 16 miles   e. 16 miles
   b. 48 miles   f. 70 miles
   c. 48 miles   g. 70 miles
   d. 16 miles

2. 76 miles + or – 4
3. 68 miles + or – 4
4. 96 miles + or – 4

1-16 FINDING YOUR WAY ON A ROAD MAP

1.a. 11   d. 13
   b. 7   e. 17
   c. 21

2. County 7, US 2, State 8, County 5, County 9
3. Forest City
4. Peachtree

1-17 DIRECTION AND DISTANCE AROUND WILDWOOD LAKE

1.a. 40 miles + or –2
   b. 30 miles + or –2
   c. 50 miles + or –2
   d. 30 miles + or –2
   e. 100 miles + or –4

2.a. NE
   b. NW
   c. E
3.a. SW
   b. S
   c. W

4. 260 miles + or – 4
5.a. Kappa
   b. Sumo
   c. Moot
   d. Delta
6. 100 miles + or – 4

1-18 USING AN INDEX TO LOCATE PLACES

1.

Community	Square	Community	Square
Mount Morris	D1	Clearfield	B3
Altoona	C3	Coudersport	A4
Williamsport	B5	Gettysburg	D4
Bloomsburg	B5	Philadelphia	D7
Stroudsburg	B7	Scranton	B6
Harrisburg	C5	Somerset	C2

2.

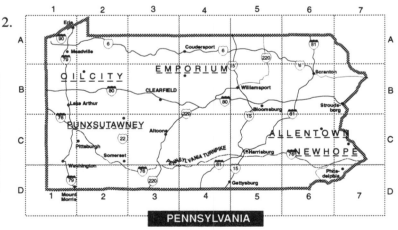

PENNSYLVANIA

3. D3, C3, C4, B4, B5, A5 (South to North)
4. B7, B6, A6, A5, A4, A3, A2, A1 (East to West)

SECTION 2: DEVELOPING A SENSE OF PLACE

2-4   CROSSWORD PUZZLE OF THE UNITED STATES II

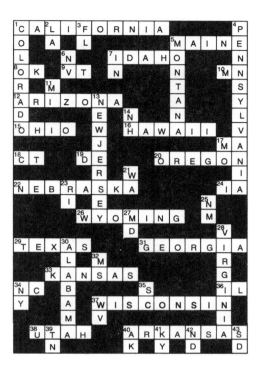

2-6   THE MISSISSIPPI VALLEY

2.a. Canadian River
   b Platte River, or Yellowstone River, or Milk River
   c. Pecos River
   d. Big Horn River
3.a. Wichita
   b. Bismarck
   c. Fort Worth

Across

Augusta
Concord
Boston
Lansing
Raleigh
Montpelier
Trenton
Jackson
Charleston
Nashville
Annapolis
Providence
Columbus
Springfield
Indianapolis

A	U	G	U	S	T	A	B	C	O	N	C	O	R	D
T	H	K	F	T	A	M	B	O	S	T	O	N	U	M
L	A	Y	R	N	L	A	D	L	G	J	L	M	X	A
A	R	W	A	I	L	L	P	U	U	M	V	D	R	D
N	T	H	N	Y	A	B	P	M	J	S	R	Y	R	I
T	F	M	K	V	H	A	B	B	M	D	E	T	I	S
A	O	P	F	L	A	N	S	I	N	G	Z	W	C	O
Q	R	P	O	M	S	Y	R	A	L	E	I	G	H	N
S	D	Y	R	F	S	K	F	E	M	O	Y	P	M	Z
M	O	N	T	P	E	L	I	E	R	F	G	I	O	H
O	V	W	E	R	E	T	R	E	N	T	O	N	N	A
N	E	M	K	L	N	U	I	R	D	E	B	A	D	R
T	R	T	D	I	J	A	C	K	S	O	N	P	F	R
G	H	C	H	A	R	L	E	S	T	O	N	L	F	I
O	B	H	F	X	N	A	S	H	V	I	L	L	E	S
M	A	N	N	A	P	O	L	I	S	B	N	M	J	B
E	Y	P	R	O	V	I	D	E	N	C	E	A	H	U
R	C	O	L	U	M	B	U	S	D	F	H	B	T	R
Y	D	S	P	R	I	N	G	F	I	E	L	D	Q	G
W	I	N	D	I	A	N	A	P	O	L	I	S	K	H

Down

Atlanta
Montgomery
Hartford
Dover
Frankfort
Tallahassee
Albany
Columbia
Richmond
Madison
Harrisburg

2-8   THE EASTERN STATES AND THEIR CAPITALS

Capital	State (abbr.)	Capital	State (abbr.)	Capital	State (abbr.)
Madison	WI	Augusta	ME	Lansing	MI
Albany	NY	Montpelier	VT	Concord	NH
Atlanta	GA	Boston	MA	Tallahassee	FL
Montgomery	AL	Raleigh	NC	Jackson	MS
Nashville	TN	Frankfort	KY	Indianapolis	IN
Charleston	WV	Columbia	SC	Annapolis	MD
Harrisburg	PA	Hartford	CT	Columbus	OH
Richmond	VA	Springfield	IL	Providence	RI
Trenton	NJ			Dover	DE

2-9 WORDSEARCH ON WEST-OF-THE-MISSISSIPPI RIVER CAPITALS

Across
Honolulu
Olympia
Cheyenne
Pierre
Austin
Baton Rouge
Sacramento
Little Rock
Salt Lake City
Jefferson City
Juneau
Lincoln

Down
Oklahoma City
St. Paul
Boise
Phoenix
Denver
Helena
Salem
Bismarck
Santa Fe
Des Moines
Carson City
Topeka

B	C	F	X	Q	H	O	N	O	L	U	L	U	W	Y	Z
O	L	Y	M	P	I	A	K	N	Z	S	B	S	D	C	H
K	C	M	E	H	E	W	B	D	M	A	I	A	E	A	G
L	S	P	X	O	Y	T	F	E	H	L	S	N	S	R	L
A	T	C	H	E	Y	E	N	N	E	E	M	T	■	S	V
H	■	B	M	N	H	V	M	V	L	M	A	A	M	O	I
O	P	O	P	I	E	R	R	E	G	R	■	O	N	K
M	A	I	C	X	S	Y	R	R	N	Q	C	F	I	■	Z
A	U	S	T	I	N	O	S	W	A	V	K	E	N	C	T
■	L	E	B	A	T	O	N	■	R	O	U	G	E	I	O
C	S	A	C	R	A	M	E	N	T	O	D	P	S	T	P
I	P	L	I	T	T	L	E	■	R	O	C	K	O	Y	E
T	S	A	L	T	■	L	A	K	E	■	C	I	T	Y	K
Y	J	E	F	F	E	R	S	O	N	■	C	I	T	Y	A
J	U	N	E	A	U	F	K	L	I	N	C	O	L	N	S

2-10 THE WESTERN STATES AND THEIR CAPITALS

Capital	State (abbr.)	Capital	State (abbr.)	Capital	State (abbr.)
Olympia	WA	Denver	CO	Des Moines	IA
Jefferson City	MO	Sante Fe	NM	Salem	OR
Sacramento	CA	Bismarck	ND	Little Rock	AR
Pierre	SD	Boise	ID	Baton Rouge	LA
Carson City	NV	Lincoln	NE	Honolulu	HI
Topeka	KS	Helena	MT	Juneau	AK
Cheyenne	WY	Oklahoma City	OK	Austin	TX
Salt Lake City	UT	Phoenix	AZ	St. Paul	MN

2-12 THE WORLD'S CONTINENTS AND OCEANS

1. North America, South America, Europe, Africa, Asia, Australia, Antarctica
2. Eurasia
3. Pacific Ocean, Atlantic Ocean, Arctic Ocean, Indian Ocean
4. Equator
5. Northern Hemisphere; Southern Hemisphere
6.a. F   b. T   c. T   d. T   e. F
7.a. Indian Ocean   b. Pacific Ocean
8. South America, Africa
9. Europe

3-2   SAIL THROUGH THE PANAMA CANAL

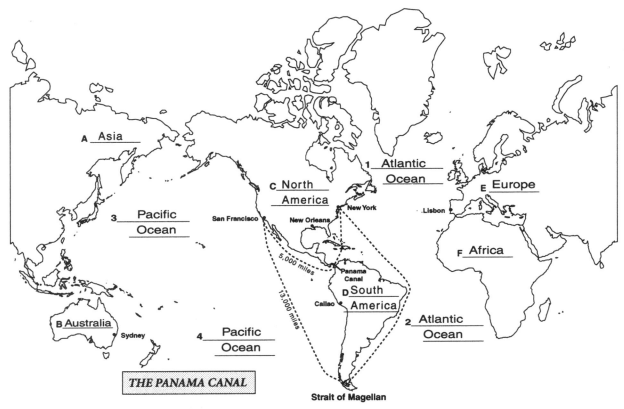

2.c. 8,000 miles
3.a. Miraflores Locks
  b. Pedro Miguel Locks
  c. Galliard Cut
  d. Gatun Lake
  e. Gatun Locks

3-4   SAIL THROUGH THE SUEZ CANAL

1.a. Southwest
  b. Cape of Good Hope
  c. Northwest
  d. 20 days
2.a. Bab el Mandeb
  b. Mediterranean Sea
  c. Strait of Gibraltar
3.   3,600 miles
4.   14.46 days (rounded to 14½ days); 5½ days, or 5.5 days (approximately)

3-6   SAIL FROM THE MEDITERRANEAN SEA TO THE BLACK SEA

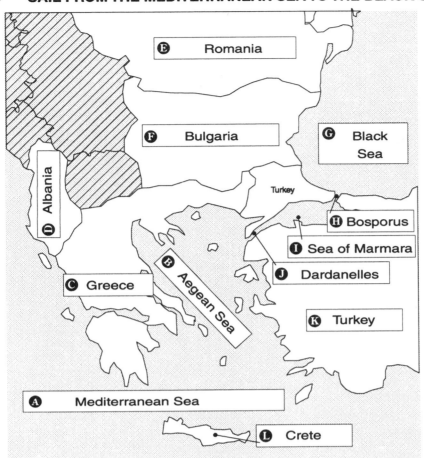

- E Romania
- F Bulgaria
- G Black Sea
- Albania
- D
- Turkey
- H Bosporus
- I Sea of Marmara
- J Dardanelles
- C Greece
- B Aegean Sea
- K Turkey
- A Mediterranean Sea
- L Crete

2. Romania and Bulgaria
3. Albania and Greece

3-8   SAIL THROUGH THE STRAIT OF GIBRALTAR

A: Denmark
B: North Sea
C: Netherlands
D: Belgium
E: England
F: London
G: English Channel
H: France
I: Bay of Biscay
J: Spain
K: Portugal
L: Lisbon
M: Strait of Gibraltar
N: Mediterranean Sea
O: Africa
P: Balearic Islands
Q: Sardinia
R: Sicily
S: Tyrrhenian Sea

3-10   SAIL THROUGH THE STRAIT OF MAGELLAN

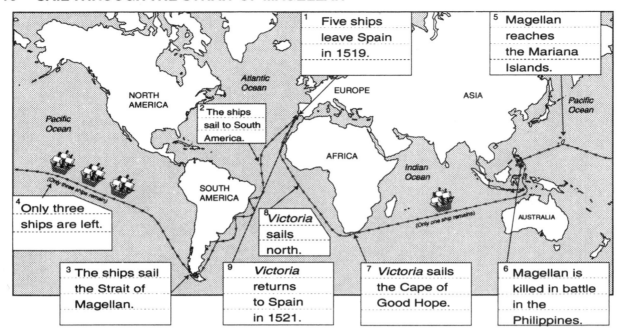

1. Five ships leave Spain in 1519.
5. Magellan reaches the Mariana Islands.
2. The ships sail to South America.
4. Only three ships are left.
8. *Victoria* sails north.
3. The ships sail the Strait of Magellan.
9. *Victoria* returns to Spain in 1521.
7. *Victoria* sails the Cape of Good Hope.
6. Magellan is killed in battle in the Philippines.

3-11   SAIL THROUGH THE BERING STRAIT

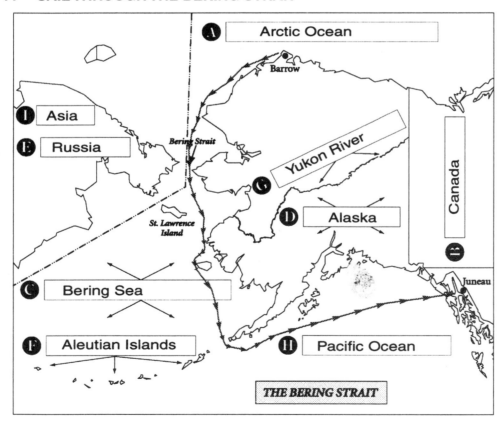

A. Arctic Ocean
I. Asia
E. Russia
G. Yukon River
D. Alaska
B. Canada
C. Bering Sea
F. Aleutian Islands
H. Pacific Ocean

THE BERING STRAIT

2.   United States

242

SECTION 4: MAPS THAT SHOW A PATTERN

4-2   TRAVELING BY ROAD AND RAILROAD

1. Zeta, Phi, Delta, Gamma, Omega
2. Sigma
3. Two
4. Omicron
5. Two
6. Beta, Omicron, Iota, Sigma, Kappa
7. 25 miles
8. $3.75 (25 miles × $.15)
9. Tago and Filo
10. Robo, Alba, Filo, Tago
11. Theta, Gamma
12. Trout River
13. Six
14. Southeast

4-3   USING TWO MAPS TO FIND ANSWERS TO QUESTIONS

1.c. (Lumberjacks)
2.b. (too many mountains and forests)
3.d. (Bauxite)
4. $6,000,000
5. $2,000,000
6.b. (($1200) (20 miles × $.20 × 300)
7.a. (Bauxite)   c. (Coal)
8.b. (Buzz)
9.d. (Tug)

4-4   A PATTERN MAP OF THE WORLD'S DESERTS

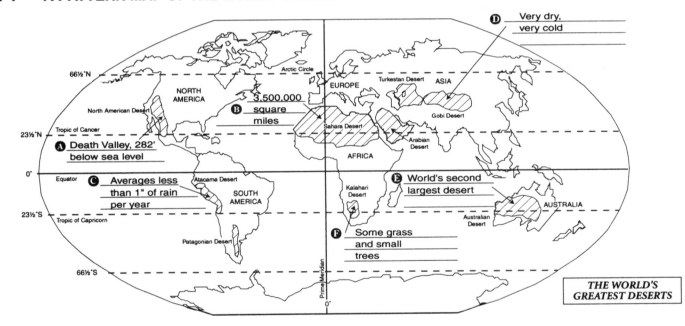

1.

CONTINENT	DESERTS
Australia	Australian Desert
Asia	1. Arabian Desert
	2. Gobi Desert
	3. Turkestan Desert
Africa	1. Sahara Desert
	2. Kalahari Desert
South America	1. Atacama Desert
	2. Patagonian Desert
North America	North American Desert

2. North of the Equator

243

4-5   AN ALTITUDE PATTERN MAP OF 19 WESTERN STATES

1.

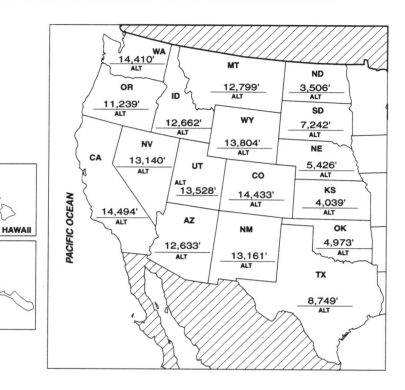

2.   10,904'

3.   Texas

4-6   A PATTERN MAP OF FOREIGN VISITORS TO THE UNITED STATES

CITY	VISITORS	CITY	VISITORS	CITY	VISITORS
New York	4,532,000	Washington, DC	1,382,000	Atlanta	566,000
Los Angeles	3,603,000	Chicago	1,110,000	Seattle	544,000
Miami	3,127,000	Boston	1,065,000	Philadelphia	431,000
San Francisco	2,923,000	San Diego	702,000	Houston	408,000
Orlando	2,583,000	Anaheim	589,000	Phoenix	408,000
Las Vegas	1,971,000	Tampa-St. Petersburg	589,000	Denver	363,000

4-7   THE LEADING BARLEY- PRODUCING STATES

1.

State	Bushels of Barley	State	Bushels of Barley	State	Bushels of Barley
North Dakota (ND)	101,250,000	Colorado (CO)	10,080,000	Virginia (VA)	5,525,000
Montana (MT)	63,600,000	California (CA)	9,900,000	Pennsylvania (PA)	5,100,000
Idaho (ID)	60,040,000	Wyoming (WY)	9,200,000	South Dakota (SD)	4,940,000
Washington (WA)	37,240,000	Oregon (OR)	8,280,000	Maryland (MD)	4,000,000
Minnesota (MN)	27,500,000	Arizona (AZ)	6,834,000	Wisconsin (WI)	3,575,000

2.   Northwest

3.   Pennsylvania, Maryland, Virginia

244

4-8 COASTLINES OF THE UNITED STATES

1.

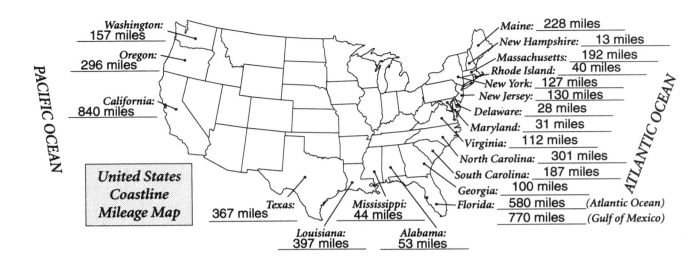

Washington: 157 miles	Maine: 228 miles
Oregon: 296 miles	New Hampshire: 13 miles
California: 840 miles	Massachusetts: 192 miles
	Rhode Island: 40 miles
	New York: 127 miles
	New Jersey: 130 miles
	Delaware: 28 miles
	Maryland: 31 miles
	Virginia: 112 miles
	North Carolina: 301 miles
	South Carolina: 187 miles
	Georgia: 100 miles
	Florida: 580 miles (Atlantic Ocean)
	770 miles (Gulf of Mexico)

United States Coastline Mileage Map

PACIFIC OCEAN

ATLANTIC OCEAN

Texas: 367 miles

Louisiana: 397 miles

Mississippi: 44 miles

Alabama: 53 miles

2. 1,350
3. California; 840 miles
4. North Carolina (301 miles); 279 miles shorter than Florida's
5. Pacific coast: 1293 miles; Atlantic coast: 2069 miles; Gulf coast: 1631 miles

SECTION 5: THE EARTH'S GRID

5-2 CONSTRUCTING A LATITUDE DIAGRAM

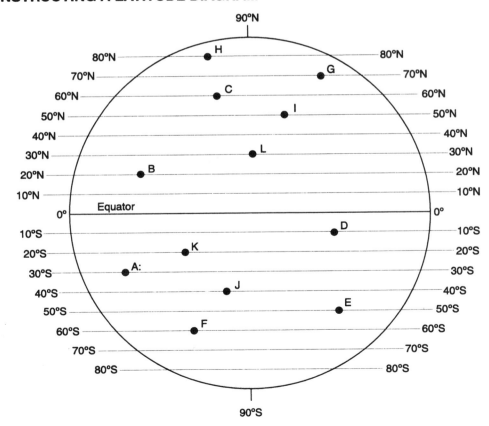

5-3 FINDING DISTANCES BETWEEN PLACES ON LINES OF LATITUDE

a. 660 miles
b. 3010 miles
c. 7155 miles
d. 5740 miles
e. 8125 miles
f. 1800 miles
g. 3864 miles

5-4 DETERMINING LATITUDE AND DISTANCE ON A WORLD MAP

PLACE	LOCATION
New York	40°N
Beijing (China)	40°N
Los Angeles	33°N (+ or − 2°)
Moscow	56°N (+ or − 2°)
A	20°S
B	20°S
C	60°N
D	20°S
E	50°S (+ or − 2°)
F	28°N (+ or − 2°)

2. South America, Africa, Australia
3.a. 3180 miles
  b. 7800 miles
4. Student choice

5-6 UNDERSTANDING THE WORLD'S CLIMATE ZONES

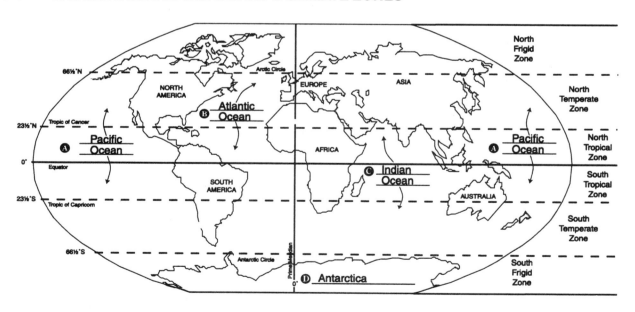

3. 66½°S
4. 66½°N
5. 23½°N and 66½°N
6. 23½°S and 66½°S
7. 23½°N and 23½°S
8. 0° (Equator)
9. Australia and Antarctica
10. North Temperate Zone

5-8 LOCATING PLACES AND MEASURING DISTANCE ON LINES OF LONGITUDE

Place	Longitude	Place	Longitude
C	45°W	G	45°W
D	15°E	H	37°W 38°W
E	60°W	I	90°W
F	52°E 53°E	J	45°W

1. Place C: 60° (4200 miles)
   Place D: 48° (3360 miles)
   Place E: 45° (3150 miles)
2. a. F and G: 5600 miles
   b. H and I: 700 miles

5-9 USING LATITUDE AND LONGITUDE TO FIND LOCATIONS AND DISTANCES

Place	Location	Place	Location
C	40°N–45°E	G	30°S–45°E
D	30°N–150°E	H	40°S–150°E
E	5°S–135°W	I	10°S–30°E

2. a. 4900 miles
   b. 4900 miles
3. H: 2800 miles
   I: 700 miles
   J: 3500 miles
4. 7200 miles

5-10 LONGITUDE HELPS US TO TELL TIME

Place	Time	Place	Time
London	12:00 Noon	Irkutsk (Russia)	7:00 PM
New York	7:00 AM	Osaka (Japan)	9:00 PM
San Francisco	4:00 AM		

a. San Francisco - New York: 3 hours
b. New York - London: 5 hours
c. London - San Francisco: 8 hours
d. London - Irkutsk: 7 hours
e. London - Osaka: 9 hours
f. Irkutsk - Osaka: 2 hours

SECTION 6: UNDERSTANDING ALTITUDE

6-2 UNDERSTANDING ALTITUDE

1. a. Hawk Plateau: 2000'
   b. Eagle Lookout: 5000'
   c. Sky Peak: 8000'
   d. Hiker's Rest: 6000'
   e. Flat Top: 6000'
   f. Dome Island: 1000'
2. a. Fish Lake: –2000'
   b. Valley Lake: –1000'
   c. Deep Trench: –4000'

3. 2000'
4. 4000'
5. 2000'
6. 4000' and 6000'
7. 7000'
8. 8000'

6-4 COMPLETING AND UNDERSTANDING A COLOR-ELEVATION MAP

1. 4000' – 6000'
2. Southwest
3. Fish River
4. 7,500'

5.a. Road B
   b. The road ends, and a trail begins.
6. Depression

6-6 CLIMBING BELL MOUNTAIN

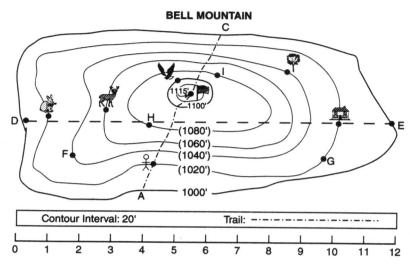

BELL MOUNTAIN

Symbol or Object		Elevation
a. House		1,020'
b. Tree		1,040'
c. Eagle		1,100'
d. Deer		1,060'
e. Rabbit		1,020'
f. Flag		1,115'

Contour Interval: 20'   Trail: ─ ·· ─ ·· ─ ·· ─ ··

0 1 2 3 4 5 6 7 8 9 10 11 12

3. C to Flag
4. C to Flag
5.a. 11½ miles (Approx.)
   b. 6 miles (Approx.)

6. A person at H would not see a person at I. There is a high hill between them.
7. 95'
8. See answer contour map.

6-7 READING A UNITED STATES TOPOGRAPHICAL MAP

1.a. 300'
  b. No
2. 354'
3. Church
4. 435'

5.a. Peters Brook
   b. Light duty
   c. Barns and a house
   d. 360'
6. 392'

Note: Check to determine if students have labeled the 380' contour line.

7-2 **GLOBE MAPS AND LINES OF LATITUDE**

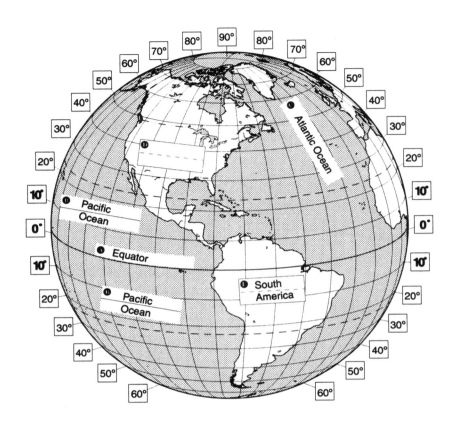

7-3 **WORKING WITH LONGITUDE I**

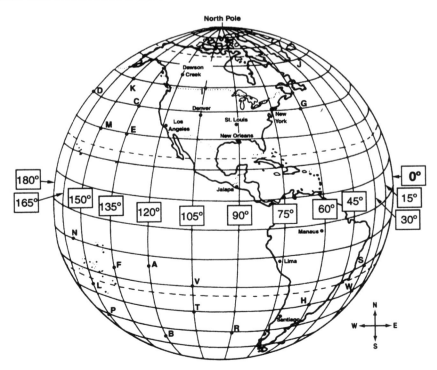

7-4 WORKING WITH LONGITUDE II

2.a. K:M, I:V, G:H, E:F, M:L, S:W

2.b. G:C, H:T, I:K, R:P, E:M, C:D

3.

PLACE	LONGITUDE	PLACE	LONGITUDE
A	120°W	G	60°W
B	120°W	H	60°W
C	135°W	I	105°W
D	180°W	J	30°W

CITY	LONGITUDE
Dawson Creek	120°W
New York	75°W
St. Louis	90°W
Los Angeles	120°W
New Orleans	90°W
Denver	105°W
Jalapa	90°W
Manaus	60°W
Santiago	75°W
Lima	75°W

7-5 LATITUDE AND LONGITUDE ON A GLOBE MAP

PLACE	LATITUDE/ LONGITUDE	PLACE	LATITUDE/ LONGITUDE
New York	40°N-75°W	A	60°N-30°W
Naples	40°N-15°E	B	50°N-45°W
Cairo	30°N-30°E	C	40°N-15°W
Leningrad	60°N-30°E	D	20°N-15°E
Ankara	40°N-30°E	E	10°N-30°W
Baghdad	33°N-45°E	F	0°-15°W
Aden	14°N-45°E	G	10°N-30°E
Durban	30°S-30°E	H	20°S-30°W
Accra	7°N-0°	I	20°S-30°E

7-7 WORKING WITH POLAR MAPS I

2.a. New York: 80°W

b. Moscow: 40°E

c. Beijing: 120°E

4.a. New York: 40°N

b. Moscow: 60°N

c. Beijing: 40°N

5.c. 3366 miles

6. A:40°N–160°E; B: 40°N–20°W; C: 40°N–140°W; D: 40°N–40°W

7. Airplane E: East; Airplane F: South; Airplane G: East; Airplane H: West

8.a. 80°E

8.b. 60°W

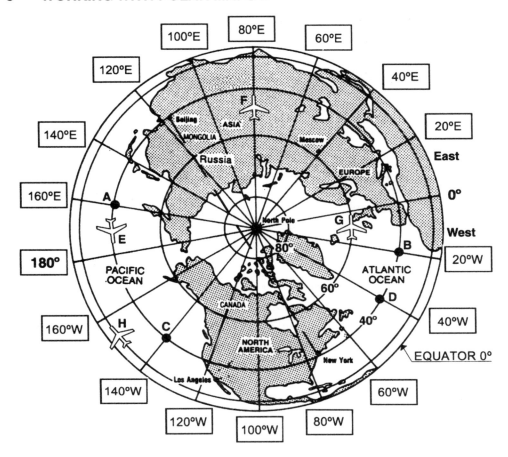

SECTION 8: GRAPHS

8-2   A PICTURE GRAPH SHOWS A FARMER'S APPLE HARVEST

YEAR	BUSHELS	YEAR	BUSHELS
2000	200	1995	200
1999	160	1994	140
1998	220	1993	200
1997	160	1992	180
1996	200	1991	160

2. 1998
3. 1994
4. 80
5. 1820
6. 8000 lbs.
7. $4,000

8. - Lack of rain
    - Late spring
    - Early freezing
    - Insect infestation
    - Failure to properly prune
    and spray
    - Severe windstorm blew
    limbs and apples off
    the trees

8-3   A PICTURE GRAPH OF A FAMILY'S JAM-MAKING ACTIVITIES

Title: **Jars of Jam Made**

Year														
1999	🫙	🫙	🫙	🫙	🫙	🫙	🫙	🫙	🫙	🫙	🫙	🫙	🫙	
1998	🫙	🫙	🫙	🫙	🫙	🫙	🫙	🫙	🫙	🫙	🫙	🫙		
1997	🫙	🫙	🫙	🫙	🫙	🫙	🫙	🫙	🫙	🫙	🫙			
1996	🫙	🫙	🫙	🫙	🫙	🫙	🫙	🫙	🫙	🫙	🫙	🫙		
1995	🫙	🫙	🫙	🫙	🫙	🫙	🫙	🫙	🫙	🫙				
1994	🫙	🫙	🫙	🫙	🫙	🫙	🫙	🫙						

JARS OF PEACH JAM

Each symbol (🫙) equals __5__ jars of jam

8-4   A PICTURE GRAPH SHOWING COTTON PRODUCTION IN THE UNITED STATES

Bales of Cotton Produced in Nine Leading States[*]

STATE										
Texas	🐚	🐚	🐚	🐚	🐚	🐚	🐚	🐚	🐚	🐚
California	🐚	🐚	🐚	🐚						
Georgia	🐚	🐚	🐚	🐚						
Mississippi	🐚	🐚	🐚	🐚						
Arkansas	🐚	🐚	🐚							
Louisiana	🐚	🐚								
North Carolina	🐚	🐚								
Arizona	🐚	🐚								
Tennessee	🐚									

- Each symbol (🐚) represents 500,000 (1/2 million) bales of cotton.
- Each bale of cotton weighs 480 pounds.
- The number of bales in each state has been rounded to the nearest 500,000.

* *World Almanac,* 1999

1. Texas, 5; Georgia, 2; Arkansas, 1½; California, 2; Mississippi, 2; Louisiana, 1;
    North Carolina, 1
2. Five times greater
3. See graph
4.a. Texas, Louisiana, Mississippi, Alabama
   b. North Carolina
   c. California
   d. Arkansas, Tennessee

8-6   THE UNITED STATES' GREATEST OIL-PRODUCING STATES

3. No   4. 150   5. 1691   6. Yes

8-7   COMPARING THE HIGHEST POINTS OF CONTINENTS

THE HIGHEST POINT ON EACH CONTINENT

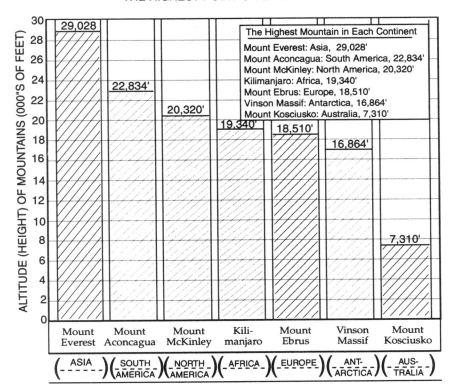

2. 3566'
3. 13,010'
4. 1340'

8-9   COMPARING QUANTITIES AND SIZES ON CIRCLE GRAPHS

Title: _The World's Six Leading_

Wheat Producing Countries

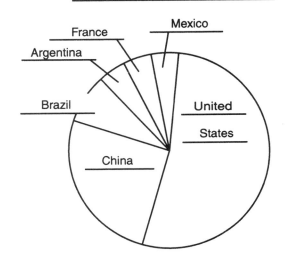

GULF STATES SIZE COMPARISON

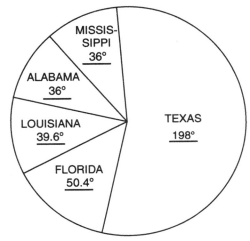

253

8-11 A SQUARE GRAPH COMPARES SOUTH AMERICA'S COUNTRIES

Country	Percent of South America
Brazil	48%
Argentina	16%
Peru	7%
Colombia	7%
Bolivia	5%
Venezuela	5%
Chile	4%
Paraguay	2%
Ecuador	2%
Guyana	1%
Uruguay	1%
Suriname	1%
French Guiana*	1%

* French Guiana is a colony of France.

2. Brazil is:
   - 32% larger than Argentina
   - 41% larger than Peru
   - 46% larger than Paraguay
   - 47% larger than Uruguay
3. No
4. Yes

8-12 TWO SINGLE-BAR GRAPHS: THE GREAT LAKES AND FISHING

Great Lakes
Percent of the whole bar: Superior, 34%; Huron, 24%; Michigan, 24%; Erie, 10%; Ontario, 8%
Fish Catch

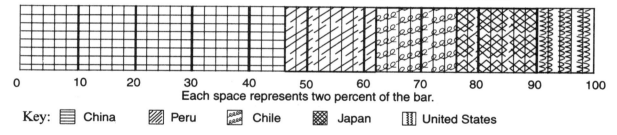

FISH-CATCH OF THE FIVE LEADING FISHING COUNTRIES

Each space represents two percent of the bar.

Key: China Peru Chile Japan United States

8-14 UNDERSTANDING TORNADOES

1. 1987
2. 1992
3. 1990 and 1991
4. 1992 and 1994

Year	Tornadoes	Year	Tornadoes
1986	764	1992	1298
1987	656	1993	1176
1988	702	1994	1082
1989	856	1995	1235
1990	1133	1996	1170
1991	1132		

8-15 COMPLETING A LINE GRAPH OF POPULATION GROWTH

TITLE: Population Growth of the United States from 1940 to 2000

5. 132 million
6. 179 million
7. 249 million
8. 275 million

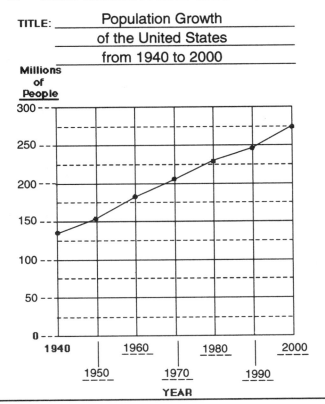

SECTION 9: SOIL EROSION AND PREVENTION OF EROSION

9-2   WHAT HAPPENED TO RIVERSIDE? I

1.a. The river would furnish transportation, water, and food.
  b. The town could be easily defended against Indian attacks.
  c. Land to the north could be easily farmed.
  d. Trees would furnish fuel and lumber.
2. Industry polluted the air with uncontrolled smoke emissions.
3. Paragraph 6: Sewage treatment
4. Paragraph 4: Polluted river
5. To be circled: 2nd, 3rd, and 4th sentences in paragraph 7.
6. To be circled: 4th and 5th sentences of paragraph 8.

9-3   WHAT HAPPENED TO RIVERSIDE? II

7. 1. Industrial waste is being dumped into the river
   2. Soil is washing into the river from a bare cliff.
   3. Garbage and "junk" are being carelessly dumped.
   4. Sewage is being dumped into the river.
8. The sand bar will make the river, and especially its channel, impassable for large boats.

9-5 SOIL EROSION: CONSTRUCTION, ROADSIDES, AND STRIP-MINING II

Construction
1. Suggested caption: Uncovered land at construction sites causes erosion.
2. Suggested: Cover the uncovered land with straw to help retard erosion; then replant.
3. Most of the loose soil will be washed into sewers and/or streams and rivers.
4. The construction has been going on for some time as evidenced by three framed houses and the severely eroded banks.

Roadsides
1. The uncovered banks will erode and will eventually undermine the guard rail and road.
2. A ditch could be planted with vegetation and/or it could be paved with asphalt, stone, or brick.
3. Protection could be offered by planting grass. Also, a ditch could be made on the side of the road to channel run-off water. Retaining walls could be constructed, and the sharp bank could have been sloped and planted.
4. A vehicle could hit the rail, the rail could give way, and the vehicle would go over the embankment.

Strip Mining
1. Strip mining leaves ground uncovered, which allows water to erode soil.
2. Many heavy trucks and earth-moving machines churn up the soil. This leaves the soil vulnerable to water and wind erosion.
3. As soon as an area has been stripped it should be covered with topsoil and planted with grass.

9-7 CAUSES OF SOIL EROSION: OVERGRAZING, BARE FIELDS II

Overgrazing
Paragraph 1: Cut, twist, and turn
Paragraph 2: Grass is eaten down to the roots. This removes protective covering from the soil
Paragraph 3: 1. Rain falls, snow melts, wind blows
           2. Paragraph 3
           3. The 2nd and 3rd sentences of the fourth paragraph

Bare Fields
1. The 2nd sentence of the 1st paragraph
2. Small rills are gradually enlarged as water runs through them. The sides and bottom erode away, and rills erode into other rills and enlarge the gullies.
3. Check student drawings.

Learning from Pictures
Picture 1
1. The gully is about twice as deep as the man is tall.
2. The depth indicates the gully has been forming for a long time. It also has weeds growing in it.
3 The gully could be filled and planted. Or, the sides could be sloped and planted.
4. This is what happens when small rills are left unattended.
Picture 2:
1. Suggested caption: Tiny rills grow to be deep gullies.
2. The twisting, turning, and cutting hooves loosen soil.
3. If left untended the rills will widen and join with other rills when winter snows melt.
4. Check student drawings.

9-11 WAYS OF PROTECTING SOIL II

1.a. 1st sentence under "Contour Farming"
   b. 2nd sentence under "Contour Farming"
   c. 2nd sentence under "Terracing"
   d. 1st sentence of 2nd paragraph of "Willow Planting"
   e. 2nd sentence of 2nd paragraph of "Wind Breaking"
2. Contours slow the flow of water downhill. They also hold water long enough for it to sink into the ground.
3. The cutting action of rivers is greatest on curves.
4. Bushes, branches, and leaves lessen the force of raindrops before they hit the ground and, thus, inhibit erosion. Also, the roots of the plants help hold soil.
5. Suggested captions:
   - Contour farming lessens the power of running water.
   - Terracing decreases the power of flowing water to erode soil.
   - Protective covering breaks the force of falling raindrops and the roots hold the soil firm.
   - Willow planting on stream and river curves reduces the power of running water to wear away banks.
6. 1. Contouring; 2. Wind breaking; 3. Willow planting; 4. Terracing; 5. Protective covering

SECTION 10: WATER: ITS SOURCES AND ITS POLLUTION

10-4 UNDERSTANDING WATERSHEDS

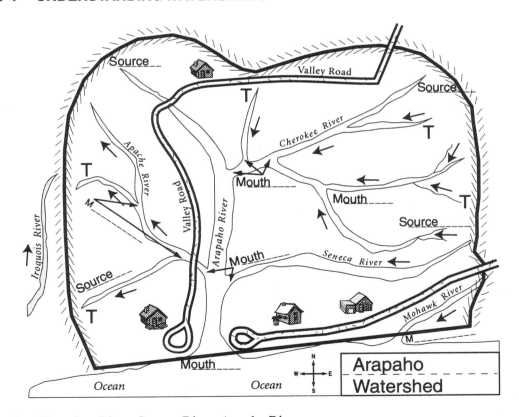

3. Cherokee River, Seneca River, Apache River
9. Outside
10. Mohawk River
11. A tree with branches

10-6   MORE ABOUT WATERSHEDS II

Picture 1
1. cover, harvest, eroded, rivers, streams, gullies, rains, winds, fewer, less, clothing, electricity
2. Suggested: deep gully, heavy rain, unproductive farm

Picture 2
1. house, barn, silo, topsoil, abandon
2. Suggested captions:
   Picture 1: Rains are causing gullies to form and soil to erode.
   Picture 2: The family leaves because they can no longer make a living on the farm.
3. Suggested: We can't fish or swim here; the water is too polluted.

10-8   UNDERSTANDING THE WATER CYCLE

1. Evaporation from: ship and tractor exhaust, animals, trees, lake, vegetation, plowed soil
2. Check student diagrams for CLOUD FORMATION
3. Rain, snow, sleet/hail
4. Run-off is flowing into the lake and the ocean.
5., 6. Check student diagrams for water storage (∘•ᵒ•ᵖ) and rock layer (//////).

10-10   WATER FROM UNDERGROUND SOURCES II

1.a. Dug well; b. Windmill; c. Artesian well; d. Drilled well
2. A bucket is lowered into the well and then raised.
3. The windmill operates only when the wind is blowing.
4. The rock layer prevents water from sinking further.
5. A drought could lower the water table.
6. The clay layer prevents water from rising.
7. Check for water droplets (∘ ° ∘ °) in the saturated layer.

10-12   SOURCES OF FRESH WATER: SURFACE II

1. Check student papers for proper labeling.
2. Water sources: rivers and lakes
3. Screening prevents objects from entering.
4. Chemicals eliminate odors and harmful bacteria.
5. A filtering tank removes fine materials.
6. Chlorine kills any remaining bacteria.
7. The weight (pressure) of the water from above forces water into mains and house pipes.

10-14   WATER POLLUTION: SPILLED OIL II

Oil-Polluting Incidents Reported In and Around United States Waters 1989–1993

Year		
1993		9,500
1992		9,000
1991		10,500
1990		10,000
1989		8,500

Number of Incidents (in 1,000's)
(Rounded to nearest 500)

**Amount of Oil Reported Spilled In and Around United States Waters
1989–1993**

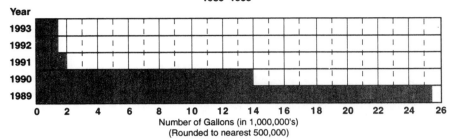

Number of Gallons (in 1,000,000's)
(Rounded to nearest 500,000)

Title: **Water Oil-Spill Incidents In and Around**

United States Waters: 1989 to 1993

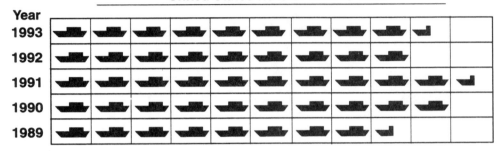

Key: Each symbol ( ) represent 1,000 oil-spill incidents

10-15 OIL SPILL IN THE SHETLANDS: AN ENVIRONMENTAL DISASTER I

1. See map
2.a. December 3, 1992
  b. Northern coast of Spain
  c. 23 million gallons
  d. 50 miles of coast covered with oil
  e. Killed and injured thousands of birds and fish
3. See map
4.a. North of Scotland
  b. Tanker struck rocks
  c. 25 million gallons
5. The force of the crash might not penetrate the second hull.
6.a. Winds: 80 m.p.h.
  b. Drinking water was contaminated
  c. ... the grass was oil-covered
  d. ... the waters from which they come are known for their cleanliness and purity
  e. A colony of seals was in serious danger from the oil that covered them and the food they needed to survive

10-16 OIL SPILL IN THE SHETLANDS: AN ENVIRONMENTAL DISASTER II

7. See map
8. There was the danger that the drift would carry the oil northeast toward Norway.
9.a. 60°N-0°
  b. See map
  c. See map

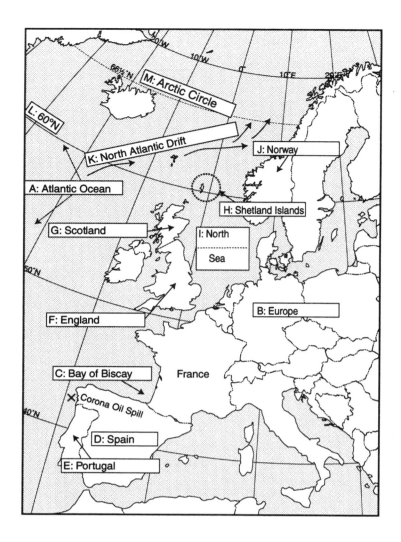

10-17 THERMAL POLLUTION: WHAT IS IT? HOW CAN IT BE CONTROLLED?

1.a. 1st sentence, 2nd paragraph
 b. Last sentence, 3rd paragraph
 c. 4th sentence, 4th paragraph
 d. 2nd sentence, 5th paragraph

3.   Discharge hot water into man-made basins. After the water has cooled, pump it into streams and rivers.

SECTION 11: AIR: WHAT IS IT? HOW IS IT POLLUTED?

11-2   ALL ABOUT AIR II

1.a. F	c. F	e. F	g. T
b. T	d. T	f. F	h. F

2.   Particles in air: pollen, microbes, salt, dust

3.   less

4.

hot-air heater

5.

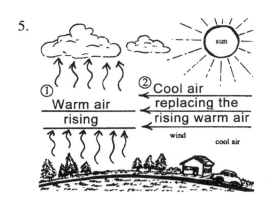

① Warm air rising ② Cool air replacing the rising warm air

sun

wind   cool air

11-4 UNDERSTANDING THE ATMOSPHERE

1. Five layers
2. Closest layer: Troposphere
   2nd closest: Stratosphere
   3rd closest: Mesosphere
   4th closest: Ionosphere
   5th closest: Exosphere
3. Mt. Everest: Troposphere
4. Troposphere
5. Stratosphere
6.a. About 3½°F
   b. 62½°
7. About 400 miles
8. Particles from activities on Earth fill the troposphere; e.g., smoke particles, agricultural dust and pollen, dust from mining, venicle exhaust, etc.

11-6 WHAT IS AIR POLLUTION? II

1. Gases, water vapor, solid particles
2. Water vapor cools and then condenses to small droplets. The droplets cling to particles in the air.
3. Air pollution comes about as a result of too many particles and strange gases entering the air.
4. Nature pollutes air as a result of forest fires, dust storms, and volcanic eruptions.
5. Volcanic ash fertilizes soil.
6. Ashes and charcoal enter the earth and enrich it.
7. Humans pollute concentrated areas for long periods of time.
8. Humans send both gases and solids into the air.
9. **Pictures**
   - Top left: Burning leaves
   - Bottom left: Fireplace and furnace smoke
   - Top right: Picnic smoke
   - Bottom right: Outdoor barbecue smoke

11-8 ACID LAKES: WHAT CAN BE DONE ABOUT THEM?

Questions
1. 45,260 gallons
2. Cities have more pollutants from vehicles, factories, etc., entering the air.
3. As rain falls it collects acids from the air.
4. Gasoline-powered motor boats. Already acidic rivers and streams could feed into a pure lake.
5. Neutralize, eliminate
6. Lakes could be sprayed from boats.
7. Fish recover from the effects of the acid, and the eggs they lay are no longer contaminated by poison acid.
8.a. Limestone dissolves easily.
   b. Limestone is inexpensive.
   c. Limestone is easy to handle.
   d. Limestone is nontoxic (not poisonous).

11-9, 11-10 AIR POLLUTION: INDUSTRY AND POWER PLANTS I AND II

1. C: Manufacturing and the Need for Power
   A: Producing Power
2. - Manufacturing plants: more than 310,0000
   - They produce the food, clothing, and the things necessary to carry on life as we know it.
   - Power without smoke: hydroelectric power, nuclear energy
   - Two most common fuels: coal and oil
   - Tons of pollutants: 43 million tons
3. Several states and communities now strictly regulate the grade of fuel used by industry.
4. The extra oxygen greatly increases the temperature of the fire and makes the flames smokeless.
5. Fly-ash is highly resistant to burning.
6. These devices cause fly-ash to stick to plates as the particles are being blown toward chimney openings.
7. Fly-ash can be used as an ingredient in cinder blocks and concrete.
8. Smoke from factories pollutes air.

Challenge: Some sources of smokeless energy: wind for windmills, solar (sun) energy, and water power.

11-11, 11-12 POLLUTED AIR IS EXPENSIVE II

1.a. Polluted air causes accidents . . .
   b. Polluted air kills cattle . . .
   c. Polluted air eats away wire insulation . . .
   d. Polluted air soils clothing, rugs, and drapes . . .
   e. Polluted air breaks down rubber tires . . .
   f. Polluted air turns paper brittle . . .
2.a. . . . erodes stone and metal statues
   b. Check student responses.
3. Crop damage: at least $500 million yearly
4. Cost to clean building: $48,000
5. Pollution cost to family of four: $260
6. Underlined phrases: kills trees; kills plants and flowers; spoils fruit (Also accept: kills cattle; cuts down on sunlight)

11-13 A POLLUTED AIR DISASTER

1. Check student diagrams.
2. Mountains (Prevent air from moving sideways)
3. Contributors to the smog: coal-burning stoves, fireplaces, factory chimneys
4. Hardest hit by smog: The very old, the very young, and those with lung-related diseases
5. Fourth paragraph beginning: The kind of tragedy . . .
6. A layer of warm air traps pollutants that arise from human activities.
7. Suggestions:
   • Face masks to filter breathed air
   • Windows and doors tightly shut
   • Factories shut down until inversion departs
   • Vehicles used only for emergencies
   • No burning of leaves, garbage, etc.

SECTION 12: WILDLIFE IN GEOGRAPHY AND THE ENVIRONMENT

12-2 WILDLIFE IN THE NEWS: EAGLES II

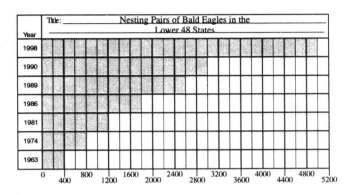

Title: Nesting Pairs of Bald Eagles in the Lower 48 States

Year													
1998													
1990													
1989													
1986													
1981													
1974													
1963													

0   400   800   1200   1600   2000   2400   2800   3200   3600   4000   4400   4800   5200

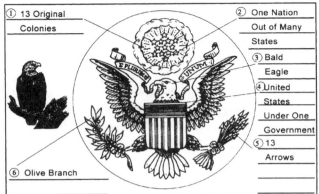

① 13 Original Colonies
② One Nation Out of Many States
③ Bald Eagle
④ United States Under One Government
⑤ 13 Arrows
⑥ Olive Branch

12-3 TREES: WHAT WE GET FROM THEM I

THE FIVE MOST VISITED NATIONAL PARKS AND RECREATIONAL AREAS

Parks

Golden Gate	☺	☺	☺	☺	☺	☺	☺	☺	☺	☺	☺	☺	☺	☺	☺	☺	☺	☺
National Capital	☺	☺	☺	☺	☺	☺	☺	☺	☺									
Lake Mead	☺	☺	☺	☺	☺	☺	☺	☺										
Great Smoky	☺	☺	☺	☺	☺	☺	☺	☺										
Acadia	☺	☺	☺	☺	☺													

☺ = 1 million visits

12-4 TREES: WHAT WE GET FROM THEM II

Note: See page 154 for list of tree products.

12-6 REDWOOD TREES MAKE A COMEBACK II

1) Pacific Ocean
2) San Francisco
3) Coast Range
4) Cascade Range

2. Tree trunk 1: 9 years old
   Tree trunk 2: 9 years old
3. Redwood is attractive, long-lasting, easily worked, free of knots
4.a. Widespreading roots offer wide support.
   b. Widespreading roots have access to water.

263

c. Thick bark offers protection from insects and fire.

5. **Redwood protection**
   a. Close regulation of cutting
   b. Protected by environment groups, governments
   c. Forests of redwood set off as national parks
   d. Redwood seedlings grown in nurseries

6. **Underlining**
   a. Gallons of water: up to 8000 gallons
   b. Thickness of bark: may be 12" thick.
   c. Age of oldest trees: 1500 or more years old.
   d. Forest fires: rough bark . . . is fire resistant; wet woods; very little underbrush
   e. Diameter and circumference: 12' diameter, 38' circumference

7. 370' equals about 37 ten-foot stories of a building.

12-8 "SPACE" AND HOW IT AFFECTS ANIMAL POPULATIONS II

- This activity is to be conducted by the instructor.
- The students' placing of objects on the squares will have to be closely observed.
- Item g, the conclusion, should be reached by the class, if possible.

12-10 ALL ABOUT WHALES II

Rorqual (Baleen Whale)

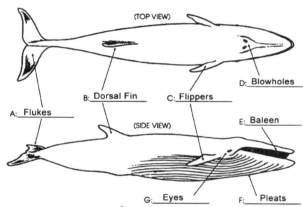

(TOP VIEW)

D: Blowholes

B: Dorsal Fin   C: Flippers

A: Flukes

(SIDE VIEW)   E: Baleen

G: Eyes   F: Pleats

12-11 WILDLIFE WORD SEARCH

C	A	R	N	I	V	O	R	E	B	D
W	O	L	F	H	S	M	T	N	A	B
F	P	S	O	W	E	N	E	D	C	A
O	R	N	X	H	A	I	E	A	T	T
O	E	A	S	A	L	V	W	N	E	G
D	D	K	T	L	I	O	N	G	R	R
C	A	E	B	E	A	R	Z	E	I	A
H	T	Q	U	L	Z	E	B	R	A	B
A	O	M	I	G	R	A	T	E	E	B
I	R	E	A	G	L	E	H	D	E	I
N	B	T	F	E	X	T	I	N	C	T
H	A	B	I	T	A	T	M	O	L	E
I	S	P	E	C	I	E	S	A	N	O
V	E	R	T	E	B	R	A	T	E	S

264

13-2 ALL ABOUT EARTHQUAKES II

1. - Number of plates: One dozen or so. (*Note*: The number depends on how they are counted.)
   - Thickness of plates: 50 to 70 miles thick
   - Where plates float: On Earth's mantle
2. Three main parts: Crust, Mantle, Core
3.a. Number of plates shown: Two
  b. Thickest: Core
4. Direction of movement: Up and down
5. The house would fall into the opening.
6. Direction of movement: Sideways
7.a. Kind of movement: Sideways
  b. Results of movement: House, barn, and fence are repositioned
  c. They would be torn apart.
8. Trench depression: Diagram 3, Movement apart

13-4 THE MAKING OF A VOLCANO

1.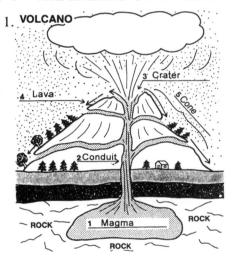

2. Magma and gases can flow out of the side vents.
3. Rock inside Earth is melted by extreme heat.
4. The magma exerts great pressure and bursts through weak spots.
5. Trees could be uprooted or covered with lava.
6. Check for coloring of the volcano diagram.

13-6 ALL ABOUT GLACIERS II

1.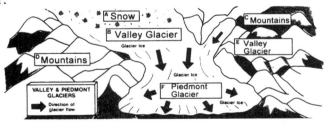

2. The weight of snow on snow presses the bottom snow into ice.
3. Glaciers move down valleys as a river might, but, of course, much more slowly.
4. The ice at the bottom of a glacier was formed years before the ice on top.
5. Piedmont glacier: 1st sentence in the 2nd paragraph
6. As the front of a glacier melts, the resultant water forms into streams and rivers.
7. As a glacier moves forward it moves and carries any loose materials in its way.
8. The materials deposited at the front of a glacier are called terminal moraines.
9. Very large glaciers are called continental glaciers.
10. Materials deposited at the sides of a valley glacier are called lateral (side) moraines.

13-7   UNDERSTANDING ICEBERGS AND ICE SHEETS

1.a. F   c. F   e. F   g. F   i. T
   b. F   d. T   f. F   h. T

2.   Most of an iceberg—some 80% to 90% of its total mass—is below the water's surface.

3.   One ice piece that broke away from an Antarctic ice sheet was much larger than Connecticut. A large iceberg can tower as much as 400 feet above the surface of the ocean.

4.   Check student horizontal and vertical lines.

13-8   UNDERSTANDING HURRICANES

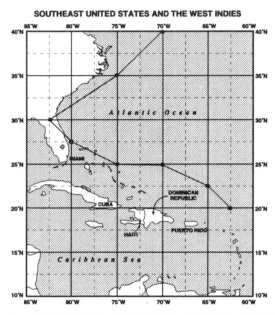

SOUTHEAST UNITED STATES AND THE WEST INDIES

1.a. June to November
   b. Hurricane Floyd
   c. Fearsome, monstrous
   d. 75 miles per hour or more
   e. Spinning top
   f. Rain, lightning, thunder
   g. At the center of a hurricane's twirling winds
   h. About 20 miles across
   i. 20 or 30 miles per hour
   j. Storm—calm—storm
   k. Eastward out to sea

13-9   GEYSERS: NATURAL WATER FOUNTAINS

1.   The order of events: 5, 2, 4, 1, 3
2.   . . .after a long period of time, minerals from the geyser's water build up into beautiful forms
3.   USA's Yellowstone National Park, New Zealand, Iceland
4.   Geysers spout and spray very hot water; people could be scalded.
5.   Check drawings.

13-10   CAVES: HOW THEY ARE MADE AND HOW THEY WERE USEFUL IN EARLY TIMES

1.   Advantages of caves: Safer, warm, food storage, weather protection
2.   Trickling water dissolves and wears away limestone.
3.   Tunnel-like
4.   Parts of cliffs are hollowed by falling water.
5.   The sea cave would be difficult to live in because, even though it may be empty of sea water at low tide, it could fill up with water at high tide.
6.   Three-quarters of a mile long, more than 600 feet wide, and almost 300 feet high.
7.

SECTION 14: LEARNING FROM PICTURES

14-2 WORKING TOGETHER IN SWITZERLAND

1. To be seen in picture: a, c, d, g, h, k, l, m, n, o, p
2.a. The cows produce milk, which can be made into cheese.
   b. The men in the picture are harvesting.
   c. The land is too steep and rocky for field crops.
   d. The men may not return from the village until the next day.
   e. It is a long climb down and up the mountain.

14-3 IN THE NETHERLANDS DIKES HOLD BACK THE OCEAN

1. A broken dike could flood the land, the houses, and the animals.
2.a. To hold back ocean water
   b. To be used as a road
3. The net is used to catch fish.
4. Meat and milk products
5. The rocks have been placed there to withstand waves and storms.
6. Steep roofs shed the rain.
7. The windmill can pump water out of the land.
8.a. The land had to be desalted.
   b. Topsoil had to be spread over the land.
9. The automobile is being driven on the left side of the road.
10. Dike: paved top; slanted land-side bank
    Cows: grazing; black and white
    Houses: slanted roofs; chimneys; stone-sided
    Windmill: four blades; stone base

14-4 A MEXICAN VILLAGE

SPANISH	ENGLISH	SPANISH	ENGLISH	SPANISH	ENGLISH
sombrero	hat	cielo	sky	niños	children
caballo	horse	gente	people	tejado	roof
mujer	woman	carro	cart	pared	wall
cacto	cactus	carpintero	carpenter	camino	road
hombre	man	ganado	cattle	árboles	trees
montaña	mountain	rueda	wheel	hierba	grass
casa	house	templo	church	herrero	blacksmith

2. Check student papers for proper circling.

14-5 ANIMAL FRIENDS AND ENEMIES IN AUSTRALIA

1. Sequence of numbering: 2, 3, 3, 4, 1, 2, 4, 3, 4, 3
2.a. To bring the sheep back into the herd.
   b. Rabbits are very numerous. They eat huge amounts of grass.
   c. The dog will herd the second sheep after he has herded the sheep he is chasing.

14-6 LIFE IN A DESERT

1. Details included in the picture: goats, rugs, poles, clumps of grass, tent, loose clothing, jugs, camel saddle, mallet
2. Four additional details: young boy holding camel; cans and pitchers; person watching sheep; bell on camel's neck
3. Statements that are probably true: a, b, e, f, g, h, i
4. Suggested main idea: Members of a desert family all work together.

14-7 FISHING IN THE UNITED STATES

1. Purpose of rock walls: To break the force of waves and currents and help keep the harbor calm.
2. Night protection: A lighthouse at the harbor entrance
3. Oncoming storm: Dark clouds and precipitation
4. Docking: Four docks
5. Cannery locations: To facilitate unloading of fishing boats
6. Open cannery: Smoke is coming from chimneys

Bottom Pictures (Titles)

1. (1) Mending nets
   (2) Drying fish
   (3) Packaging fish
2. a. Transported to stores
   b. Placed on shelves (*Note*: The fish may have to be frozen. If so, that would be the first step.)

14-8 LIFE ON THE PLAINS IN EARLY TIMES

1. Picture phrases:
   Left column, top to bottom: 5, 10, 4, 7, 6
   Right column, top to bottom: 2, 9, 8, 3, 1
2. a. Fleeing a frightening tornado (8)
   b. Farming with a steel plow (4)
   c. Stuck on the muddy road (10)
   d. Trying to stop a grass fire (5)
   e. Traveling by Conestoga wagon (1)
   f. Building a sod house (2)
   g. Digging for fresh water (3)
   h. Fighting a blinding blizzard (7)
   i. Erecting a barbed-wire fence (9)
   j. Clouds of hungry grasshoppers (6)

14-9 NATURE AND HUMANS CHANGE THE FACE OF EARTH

1. 1 (Nature), 2 (Human), 3 (Human), 4 (Nature), 5 (Nature), 6 (Human)
2. 3, 2, 1, 4, 6, 5
3. 5 (Also, picture 3 shows smoke coming from ship.)
4. 6
5. The soil was picked up and carried by the river and deposited at the river mouth.
6. The crashing, whirling waves wore away the rock.

SECTION 15: EARTH, SUN, AND MOON RELATIONSHIPS

15-1 DAY AND NIGHT: WHAT CAUSES THEM?

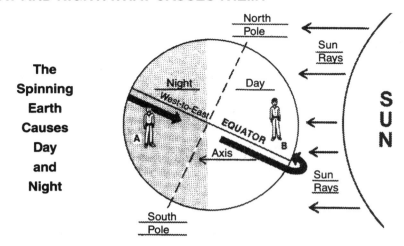

The Spinning Earth Causes Day and Night

15-3 REMEMBERING MOON FACTS

1. F	6. F	11. F	16. F
2. F	7. T	12. T	17. F
3. T	8. T	13. F	18. F
4. F	9. T	14. F	19. T
5. T	10. F	15. F	20. T

15-5 SEASONS NORTH AND SOUTH OF THE EQUATOR

1. June 22
2. North
3. Tropic of Cancer
4. hotter
5. December 22
6. hot
7. North
8. North Pole
9. warm, cool
10. D
11. The tilt of Earth causes the sun's rays to strike Earth's parts at different angles during Earth's revolution around the sun.

15-7 UNDERSTANDING THE COMPASS: DIRECTION AND TIME

1. a. South
   b. East
   c. Southwest
   d. Northwest

2.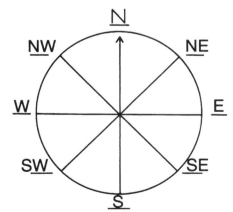

3. a. 55 minutes
   b. 35 minutes
   c. 105 minutes

16-2   CROSSWORD: COUNTRIES OF THE WORLD

1 S	L	O	2 V	A	3 K	I	4 A		5 A	L	G	E	R	I	6 A
P			E		E		U	7 J							R
A			N		N		S	A	8 C						G
I		9 E	G	Y	P	T		10 P	O	L	A	N	D		E
N		Z	A		R		A		N						N
		U			A		11 N	E	P	A	L				T
12 C	13 P	E	R	U	L		A				A		14 I		I
H	L			15 L	I	B	E	R	I	A		R		N
16 I	S	R	A	E	17 L	A				E		A
L			I			18 C	H	19 I	N	A		L
E		20 C	U	B	A			R			A
			Y		21 C	H	A	D			N
22 I	C	E	L	A	N	D		N	23 I	N	D	I	A

16-10   LATITUDE AND LONGITUDE TREASURE HUNT

1. 60°S–50°W
2. 50°S–20°W
3. 50°S–10°E
4. 30°S–60°E
5. 30°S–30°E
6. 20°S–0°
7. 10°S–30°W
8. 10°N–50°W
9. 30°N–70°W
10. 40°N–40°W
11. 50°N–20°W
12. 40°N–10°W
13. 20°N–0°
14. 10°N–10°E
15. 20°N–40°E
16. 30°N–60°E
17. 0–50°E
18. 0°–10°E
19. 10°N–30°W
20. 20°N–40°W

16-12   UNSCRAMBLE THE NAMES OF COUNTRIES AND CAPITALS

Scrambled Countries

1. India
2. Denmark
3. Hungary
4. Chile
5. Brazil
6. Sweden
7. France
8. Austria
9. Belgium
10. Canada
11. Albania
12. Spain
13. Peru
14. Turkey
15. Colombia
16. Ethiopia
17. Vietnam
18. Finland
19. Sudan
20. Bolivia

Scrambled Capitals of the United States

1. Trenton
2. Albany
3. Sacramento
4. Boston
5. Bismarck
6. Topeka
7. Augusta
8. Nashville
9. Tallahassee
10. Raleigh
11. Honolulu
12. Helena
13. Springfield
14. Cheyenne
15. Atlanta
16. Providence

16-14   ENVIRONMENTAL RIDDLES

1. - Snapdragons
    - Dragonflies
    - Dandy Lions (Dandelions)
    - Tiger Lilies
    - "Nice trunk you have there."
    - "Bye, son." (Bison)
    - Rattles
    - "I guess I'm going to be re-tired." (Retired)
    - They are both "bucks."
    - "Give me a break!"
    - "Bug off!"
    - "I get a charge out of you."

2. Suggestions
    -   How would you start a letter to a deer?
        (Dear Deer)
    -   What is the favorite song of windmills?
        ("Blowin' in the Wind")
    -   How are an alligator and a saw alike?
        (They both have lots of teeth.)

3. Suggested last line:
    Is up to you and up to me.